T5-BCI-866

ASBESTOS

ASBESTOS
ITS HUMAN COST

JOCK McCULLOCH

UNIVERSITY OF QUEENSLAND PRESS

ST LUCIA • LONDON • NEW YORK

HD
7269
.A552
A86
1986

First published 1986 by University of Queensland Press
Box 42, St Lucia, Queensland, Australia

© Jock McCulloch 1986

This book is copyright. Apart from any fair dealing for the
purposes of private study, research, criticism or review, as
permitted under the Copyright Act, no part may be reproduced
by any process without written permission. Enquiries should
be made to the publisher.

Typeset by University of Queensland Press
Printed in Australia by Dominion Press—Hedges & Bell

Distributed in the UK and Europe by University of Queensland Press
Dunhams Lane, Letchworth, Herts. SG6 1LF England

Distributed in the USA and Canada by University of Queensland Press
250 Commercial Street, Manchester, NH 03101 USA

Cataloguing in Publication Data

National Library of Australia

McCulloch, Jock, 1945–
    Asbestos–its human cost.

    Bibliography.
    Includes index.

    1. Asbestos mines and mining — Australia — Hygienic
    aspects. 2. Asbestos — Australia — Physiological effect.
    3. Asbestos industry — Australia — Hygienic aspects. I.
    Title.

363.1'79

British Library (data available)

Library of Congress

McCulloch, Jock, 1945–
    Asbestos–its human cost.

    Bibliography: p.
    Includes index.

    1. Asbestos industry — Australia — Employees — Diseases
    and hygiene. 2. Asbestos — Environmental aspects.
    3. Asbestos — Toxicology. I. Title.

HD7269.A552A86   1986   363.1'79   85-31487

ISBN 0 7022 2001 9

To Phil Thompson

University Libraries
Carnegie Mellon University
Pittsburgh, PA 15213-3890

He had alarmed her affection and her conscience by the shadowy image of consequences; he had arrested her intellect by hanging before it the idea of a hopeless complexity in affairs which defied any moral judgment.

George Eliot, *Romola*

Contents

Acknowledgments   *xi*
List of Illustrations   *xiii*
Chronology of Events   *xv*
Abbreviations   *xix*
Glossary of Terms   *xxi*

1. Introduction   *1*
2. The Industry   *8*
3. The Medical History   *36*
4. Wittenoom: The Mine   *70*
5. Wittenoom: The People   *101*
6. Baryulgil: The Mine   *131*
7. Baryulgil: The People   *159*
8. The Politics of Asbestos   *190*
9. Safety and Profits   *230*
10. Conclusion   *254*

Notes to Chapters   *266*
Selected Bibliography   *286*
Index   *293*

Acknowledgments

This book was written by accident. In January 1984 I was appointed to a visiting fellowship in the Social Justice Project, RSSS ANU. I had intended to research and write a history of Aboriginal health. That project was put aside however in favour of the story of Baryulgil which is a small and insignificant Aboriginal community. In order to understand what happened at Baryulgil it was necessary to understand the history of the asbestos industry in Australia and in particular the career of the medical literature about those diseases associated with asbestos mining and asbestos products. Out of that research grew the present book.

There are numerous people who in various ways helped in the writing of that story. I owe a particular debt to Pat Troy as head and inventor of the Social Justice Project and to James McNulty, Sidney Sax and Trevor Francis each of whom read the manuscript in its entirety. I am also indebted to Johanna Sutherland and Michaela Richards who assisted with researching the material on which this study is based. To the people of Baryulgil and to Mr and Mrs Ken Gordon I owe a special debt. I would also like to thank in no particular order of merit Chris Lawrence, Ann McDermott, Nancy Stefanovic, Carole Sutherland, Sue Dimovska, Jan Wapling, Judy Barber, Lee Mckay, Hubert Joosten, Joy Horton, Maxine Sells, the Asbestos Diseases Society, John Braithwaite, Paul Havermann, Christine Helliwell, Wade Chambers, Norma Chin, Ann McCulloch, Andrew Chin, David Anderson, Kate McCulloch, the US Information Service, Peter Karmel, Dorothy Bloom,

Lyndsay Farrall, Roxanne Blake, Brian O'Neill, Cecil Patten, Neil Walker, and Mrs Lucy Daley.

Finally, I owe special thanks to Alexander McCulloch for watching Playschool and Romper Room which made the writing of this book possible.

I cannot unfortunately thank either James Hardie Industries or CSR, both of whom refused to discuss the question of asbestos and the behaviour of their firms at Baryulgil and Wittenoom.

List of Illustrations

Wittenoom

Following page 106

1. The Wittenoom mine.
2. A section of the housing provided by ABA for its employees at Wittenoom.
3. Tailings from the mine were used to top-dress roads and paths about the miners' homes.
4. The fibre from the ABA mill was packed by hand into hessian bags and shipped by truck to the coast.
5. A willi-willi approaching a miner's home at Wittenoom.
6. The Wittenoom airport, *c.* 1955.
7. The entrance to the Wittenoom mine, *c.* 1958.
8. Working the mine face.
9. A group of miners, *c.* 1960.
10. A miner operating a scraper in the Wittenoom mine.
11. (a) and (b). Conditions inside the Wittenoom mine were cramped. The miners laboured in low shafts where dust and airborne fibre generated by the mining was inescapable.
12. A group of miners in the crib room during a break.
13. A funeral of an unknown person at Wittenoom.

Baryulgil

Following page 170

1. The mill at Baryulgil.
2. Part of the mine.

3. During the first years of the mine's operation during World War II, the host rock and tailings were carted by horse-drawn skip from the mine site.
4. A miner using a jack hammer to break down ore into workable pieces in the quarry.
5. The bagging section at the mine.
6. The mine truck was used for various purposes at the mine and mill.
7. The asbestos ore at Baryulgil was mined in an open quarry with drills and the use of high explosives.
8. The mine in 1956.
9. The Baryulgil mine in the late 1960s.
10. The mill which was demolished by Woodsreef after the closure of the mine in 1979.
11. Asbestos fibre left in damaged bags at the Baryulgil mine when the site was finally closed by Woodsreef in 1979.
12. The mine showing packed bags of fibre waiting to be trucked to the railhead at Grafton.
13. The tailings heap adjacent to The Square.
14. (a), (b), (c) and (d). The tailings heap less than one kilometre from The Square, showing the erosion by wind and water.
15. A group of Baryulgil residents including Andrew Donnelly.
16. Two Baryulgil children playing in an area which has been top-dressed with asbestos-rich tailings.

Chronology of Events

1885	The Colonial Sugar Refinery (CSR) founded.
1892	James Hardie company founded by Andrew Reid.
1898	Initial reference to the dangers posed by asbestos dust in a British Government Annual Factory Inspectors' Report.
1901	Foundation of the American asbestos conglomerate Johns-Manville.
1903	James Hardie begins to import French asbestos products.
1906	Montague Murray, a British physician, describes a case of asbestosis in evidence before a Commission on Industrial Disease.
1916	Foundation of the British asbestos producer Turner Brothers (Turner & Newall).
1919	The American state of Wisconsin enacts Workmans' Compensation legislation covering asbestosis.
1924	Publication of an article by W.E. Cooke in *The British Journal of Medicine* about an asbestos factory worker. An autopsy showed massive deposits of asbestos dust in her lungs.
1928	H.E. Seiler publishes results of a study in *The British Journal of Medicine* of autopsy results involving asbestosis.
1930	Publication of the important Merewether and Price report on the effects of asbestos dust on the lungs of British factory workers.
1932	Memorandum by the British government on silicosis

	and asbestosis. The memorandum lists dusty occupations and recommends reducing exposure levels.
1932	British physician publishes paper on asbestosis and the methods for its diagnosis.
1933	Settlement of major compensations claims in the United States involving the asbestos conglomerate Raybestos.
1933	American producer Johns-Manville settles a number of claims involving asbestosis among employees.
1934	James Hardie establishes its own research division at its Camellia plant in Sydney.
1938	American Conference of Governmental Industrial Hygienists recommends a standard maximum permissible level of exposure.
1938	Publication of Lanza's *Silicosis and Asbestosis*, which was to remain the standard work on these diseases for many years.
1942	Study published by H. Holleb and A. Angirst on asbestosis and cancer of the lung among asbestos insulation workers.
1943	Australian Blue Asbestos (ABA) formed as partnership between Lang Hancock and CSR to exploit the asbestos deposits at Wittenoom. In 1949 ABA becomes a fully-owned subsidiary of CSR.
1943–45	Eight American states enact Workers Compensation Legislation to cover the disease of asbestosis.
1944	James Hardie becomes involved in the mining of asbestos at Baryulgil. Hardie forms a partnership with Wunderlich and forms Asbestos Mines Pty Ltd (AM).
1947	Completion of Merewether study on cancer of the lung among asbestosis sufferers.
1947	New South Wales Factories and Shop Act extended to cover asbestos cement roofing manufacture.
1950	Third International Conference on Pneumoconiosis held in Sydney.
1951	Publication in *Lancet* of study by S.R. Gloyne on cancer of the lung among asbestosis sufferers.

1953   Asbestos Mines Pty Ltd becomes a fully-owned subsidiary of James Hardie.

1955   Study by Richard Doll showing high rates of cancer among British asbestos workers.

1955   Dr Heuper of the United States Department of Health publishes monograph connecting asbestosis with cancer and suggesting the existence of danger from environmental exposure.

1958   Swedish industry begins search for asbestos substitute for use in shipbuilding.

1960   Publication of the important study by Wagner showing a high incidence of mesothelioma among those coming into occupational and environmental contact with blue asbestos fibre in a South African mining community.

1961   Study by Sleggs verifying Wagner's work on mesothelioma.

1962   Australian physician James McNulty publishes one of the first clinical profiles linking exposure to asbestos with mesothelioma.

1964   Publication of Irving Selikoff's seminal research on cancer among New York insulation workers.

1964   International conference in New York on the health effects of asbestos.

1964   Publication of work by Enticknap and Smith showing an association between asbestosis, cancer of the lung, and mesothelioma.

1966   The CSR-owned subsidiary ABA closes the Wittenoom mine.

1968   First Australian Pneumoconiosis Conference held in Sydney.

1969   New British standard for exposure introduced, which set the maximum exposure level at 2 fibres per cubic centimetre. The standard was based on the prevention of asbestosis and gave no protection against mesothelioma or cancer of the lung.

1969   Asbestosis becomes a scheduled disease under the New South Wales Dust Diseases Act.

1972   International conference in Lyons on the effects of asbestos on health.

1976   Asbestos Mines Pty Ltd is sold by Hardie to Woodsreef Mines Ltd, which operate the Baryulgil mine until its closure in 1979.

1977   Death of Andrew Donnelly, a Baryulgil miner, which first alerted the workforce to the dangers of asbestos.

1978   CSR establishes the Wittenoom Trust to distribute charity to the victims of the mine.

1980   Jan Joosten dies from mesothelioma.

1981   Williams Inquiry Report into occupational health and safety tabled in New South Wales Parliament.

1982   Johns-Manville, the American asbestos conglomerate, files for voluntary bankruptcy under Federal Law.

1984   House of Representatives Inquiry report into Baryulgil released.

Abbreviations

ACOA   Australian Clerical Officers Association
ADS   Asbestos Diseases Society
ALS   Aboriginal Legal Service (Redfern)
AMS   Aboriginal Medical Service
AWU   Australian Workers Union
CTHC   Capital Territory Health Commission (ACT)
DAA   Department of Aboriginal Affairs
DDB   Dust Diseases Board of New South Wales
EPA   Environment Protection Agency (USA)
FMWU   Federated Miscellaneous Workers Union
IAC   Industries Assistance Commission
IARC   The International Agency for Research
            on Cancer
ILO   International Labour Organization
NH & MRC The National Health and Medical
            Research Council
SPAS   South Pacific Asbestos Society
WHO   World Health Organization

Glossary of Terms

Asbestosis: A form of pneumoconiosis caused by the inhalation of asbestos fibre. As it develops, the disease may cause a crippling fibrosis of the lung tissue. As the lung loses elasticity its capacity to function is reduced. The victim will suffer from progressive shortness of breath. The disease can be fatal and will not be diminished by removing the afflicted individual from the hazardous environment in which the disorder was contracted.

Carcinogen: In current usage, a carcinogen is defined as a substance capable of producing cancer.

Epidemiology: The study of the distribution of diseases (in time and locality) of persons, with the aim of discovering their history, including their response to various forms of treatment.

Exposure: Exposure to asbestos dust may occur during mining or milling or during any process involving asbestos-based products. Asbestos is not only inhaled; it can also be ingested through water and beverages such as wines and spirits.

Lung cancer: This disease is common among asbestos workers. Some studies suggest that a majority of asbestos workers who smoke heavily will develop cancer of the lung.

Mesothelioma: A rare primary cancer of the lung (pleura) or the abdominal cavity (peritoneum) which is invariably associated with exposure to asbestos. The disease is almost always fatal and life expectancy from the date of diagnosis is from one to three years. The disease is marked by the long latency period between original exposure and the appearance of the disorder. This ranges from twenty to forty years. There is no known level of exposure to asbestos at which mesothelioma cannot occur. But the disease is most commonly associated with exposure to crocidolite or blue asbestos.

Pleural plaques: These are patches of thickening which appear on the lining of the chest wall or over the diaphragm in the pleural membrane. They commonly occur before fibrosis. In some countries, plaques are accepted as an indicator of asbestos-related disease.

Pneumoconiosis: A generic term for occupational diseases of the lung. These diseases are invariably caused by the inhalation of large quantities of dust over a prolonged period of time. All forms of pneumoconiosis lead to increased susceptibility to infections such as tuberculosis, chronic bronchitis, emphysema and cancer. Asbestos dust is but one of a number of dusts, including dust from coal and silica, which cause disease.

Toxin: An agent which causes some form of damage to a living organism is said to be toxic. In theory many substances have the ability to cause damage to living organisms if administered in sufficiently high dosage. It is important, however, to note that few substances, including synthetic chemicals, have the ability to cause cancer in animals or in humans.

1

Introduction

Most Australians will come into contact with asbestos fibre at some time in their life. Contact may occur in the workplace, in the street, the train or in the home; asbestos is the most commonly used carcinogen in the contemporary world. It is a substance which has brought many advantages to producers and to consumers: asbestos fibre resists heat and cold and sound; it strengthens cement sheeting and pipes which carry water and sewage to urban populations. It also causes a number of appalling diseases which have ruined the lives of asbestos workers and the consumers of asbestos products. At least one of those diseases was well known long before asbestos was mined commercially in this country and long before asbestos products were introduced into virtually every Australian home. The stories of the asbestos mines at Wittenoom in Western Australia and Baryulgil in New South Wales help to explain how such a thing could happen.

People living in an urban environment cannot escape inhaling some asbestos fibre; no social strata is immune. In January 1980 it was revealed that asbestos insulation was to be removed from the Royal Yacht Britannia. The yacht was built in 1952, and like most vessels from that era asbestos insulation was used extensively throughout its structure. The cost of removal was not made public, but the news caused something of an outcry in Britain as much out of fear for the welfare of the Royal Family as for the millions of pounds to be spent on removal during a period of economic recession. Asbestos had already been identified as a carcinogen at the time it was used in the Royal yacht and it continued to be used widely throughout the next thirty years. On 16 February 1985, *The Canberra Times* revealed that a large number of custom-built fire doors intended for the new Parliament House had been buried at the Gungahlin tip. When the doors arrived at the building site they were found to contain asbestos. The trade union in control of the site had rejected initial reassurances that the doors were asbestos-free and after examination it was discovered that over half of the consignment contained fibre. The doors were not returned to the manufacturer for fear that they would merely be resold to another, less discriminating customer. They were trucked to the tip by the Department of Territories and buried using Department equipment. The total cost of dumping the doors was in excess of $50,000. Without trade union intervention, the new Parliament House would have been allowed to resemble the original building which is heavily insulated with asbestos where it is found in both Senate and House of Representatives extensions added after 1955.

A society may best be judged by the way in which it treats its weakest members. The Soviet Union is invariably condemned, and rightfully so, for its treatment of dissidents and those elements whose existence threatens the expansion and survival of the Soviet empire. A capitalist society such as Australia has its own victims, and in the case of asbestos they are to be found in a variety of occupations involving asbestos products. Those victims are people who have fallen ill through an industrial process dedicated to the accumulation of wealth and advantage. The very weakest members of the Australian workforce are recent immigrants and Aborigines. They are also the two groups at whose expense the wealth of this country had largely been torn from the ground. Many of the men who mined crocidolite or blue asbestos for CSR at Wittenoom were recruited in Southern Europe specifically to work in the mine. The labour force at Baryulgil was drawn almost entirely from within the local Aboriginal community which consisted of the surviving members of the Banjalang tribe. At Wittenoom and Baryulgil the social origins of the workforce, the indifference of corporate attitudes and the negligence of the regulating authorities put at risk the health of both communities.

Contemporary industry has various ugly faces, which may be turned towards employees, to those who live in the shadow of plants and factories, to those citizens who unwittingly consume hazardous products, and to those who are forced by chance to live with its wastes. The tragedies which have occurred at Minamata Bay in Japan, at Seveso in Italy, at Love Canal in New York State, and most recently at Bhopal in India each presents an element which is to be found in the asbestos story.

Individually, none of those disasters is as complex or as profound in its implications as is the story of asbestos, and what can be gleaned from those episodes is best described as a series of oppositions. Those oppositions see the weak opposed by the powerful; they see a divorce in time between action and responsibility; they see the collusion (sometimes involuntary) between commerce and government and they each testify to the limits of scientific inquiry in furthering the achievement of social justice. These episodes also reveal the way in which the legal system can be exploited for the benefit of the strong. Above all else, they tell us something of the impact of industry on the lives of the people involved, and of the weight of that material interest which is so readily arraigned against those who seek compensation for their suffering.

The asbestos story is about how the greed of the market place, the ignorance and trust of consumers, and the feebleness of governments have worked in concert to allow the destruction of the health of employees and consumers. Asbestos is not, of course, the only such case. What is important about asbestos is that it is the first disaster on such a scale, and therefore tells us something of our collective futures. Since 1945 numerous synthetic products have been marketed without adequate precautions to ensure the safety of the users of those products or of the men and women who made them. The victims of asbestos just happen to be the first generation of those crippled by hazardous substances. Asbestos is made unique only by the sheer scale of its use in the past and by the cost of rendering safe the asbestos already present in the environment.

Only in the past two years has it become apparent to what extent asbestos fibre forms part of the daily work environment of most Australians. On 21 March 1985, the National Library in Canberra was evacuated when air monitoring equipment revealed dangerously high levels of asbestos contamination. Library staff and patrons were evacuated at 2.00 p.m. on what was the fourth occasion on which the building had been declared unsafe. That building is only one of many which feature asbestos insulation. The Commonwealth alone faces a potential bill of $9 billion for the removal of asbestos from public buildings in every state, and work has already begun on a number of major sites including the Commonwealth Centre in Melbourne, which is located on the corner of La Trobe and Spring streets. Australia's leading property investor, AMP, is also faced with the huge cost of removal from many of its most prestigious holdings. Its major single property, BHP House in Bourke Street, Melbourne, is insulated with asbestos, and the company is confronted not only with the cost of removal but also the cost of relocating tenants. The largest Victorian firm specializing in removal, Bayside Asbestos Removal, estimates that over 70 per cent of all buildings in Melbourne's central business district which are more than ten years old contain asbestos. The material is so common in Collins Street that in the trade the street is usually referred to as "Asbestos Alley". The final cost involved for inner-city properties could be well in excess of $200 million.

The widespread presence of asbestos in the environment means that the asbestos story is not just about occupational health and the fate of those who have worked in asbestos cement plants or in asbestos mines. The men and women who produced asbestos products are but a minority of those who have come into contact with the fibre. People in work situations enjoy the advantage of having greater persuasive power, in being able to withdraw their labour, but consumers or those citizens such as at Bhopal in India caught in the fall out from an industrial accident have little comeback. Employees are more able to make a claim for compensation than are the users of dangerous products. The conflict between employers and employees over the effects of asbestos is important because it is the forerunner of a confrontation between communities and capital about the quality of urban life. In a society in which an increasing number of people subsist outside the workforce, this conflict is becoming more important than that between labour and capital. Communities are made up of workers and citizens, families, children, consumers, invalids, the destitute and the powerless. In all OECD states, those who are employed are becoming a shrinking majority and they have in any case the advantage of a voice, denied to many of their fellow citizens. The threat from toxic and carcinogenic substances is also present outside the factories which produce pesticides, vinyl chloride and asbestos cement sheeting. This is clear from the history of Wittenoom and Baryulgil where the families of miners were heavily exposed to fibre.

In understanding the asbestos story it is important to draw a distinction between the victims and the causalities. The victims are the miners and process workers and the consumers of asbestos products. They are the men and women whose lives and those of their families have been ruined by asbestos-related disease. It is they who, up until now, have occupied the stage. The causalities will be drawn from the ranks of any found culpable. They are the major actors who, to date, have managed to remain off-stage. In the United States the causalities have already begun to appear, and Australian producers are now acutely aware that they may well face the same fate as the bankrupt Johns-Manville corporation, the major American manufacturer of asbestos products. The two Australia firms which operated asbestos mines are Jamies Hardie and CSR, and both fear that they may come to emulate Johns-Manville's experience.

The asbestos mine at Wittenoom was operated as a subsidiary of CSR from 1944 until its closure in 1966. It took more than ten years after that date for the full impact of the exposure of the workforce to high levels of fibre to become apparent. The actual numbers of men and women who have died because of that mine may still be counted in the hundreds. But the effect of the diseases which Wittenoom has caused cannot be measured just in numerical terms. Evidence suggests that illness and distress to wives, husbands and children and the destruction of families has been common among employees of Australian Blue Asbestos. Most disturbing of all is the fact that such an outcome was predictable even from the mine's first days. The only surprising aspects of events at Wittenoom is that so few are known to have died. At Baryulgil the situation which developed allowed an isolated Aboriginal community to remain free from intrusion by state and welfare workers. Asbestos Mines Pty Ltd, a subsidiary company of James Hardie Asbestos, was the means by which the people of Baryulgil were tied to the land and to the mine.

In most contemporary industries, employees have little or no control over the work process and they understand little of the way in which products are made and why. That part of the assembly line faced by each man and woman is but an abstract fragment because the process of production has been increasingly broken down into simplified operations. That fragmentation has only increased the subordination of employees. This was not true at Baryulgil where each worker was capable of operating and servicing most if not all of the machinary and plant. That was so because the mine and mill were so small and relatively simple in design and function. The Aboriginal workforce had also played a major part in constructing the plant, and no part of the industrial process was mysterious or unknown. That knowledge, however, did not protect the employees of Asbestos Mining from asbestosis, mesothelioma and cancer of the lung, all of which are diseases associated with exposure to fibre. The CSR-owned mine at Wittenoom was far larger, and the process of mining and milling fibre was more complex. At Wittenoom, employees had less understanding of their immediate work environment. But like the men of Baryulgil they knew nothing of the known effects of fibre on the lungs. Understanding of the raison d'être of a plant or industrial process or a strong sense of solidarity among employees is not sufficient to give protection against accident or disease. The only safeguard lies in an informed and vigilent workforce and in competent and responsible regulating authorities. Both were absent at Wittenoom and Baryulgil.

Wittenoom and Baryulgil are essentially the same story told not once but twice. The first ended in 1966 and the second dragged on for a further thirteen years. In reading these stories it is difficult to understand why successive generations of miners and their families were exposed to a known hazard. The great fear now is that the same story will be told again and again among the consumers of the products from those mines. At Wittenoom and Baryulgil there were a series of elements which in each case made events possible. But there is no explanation as to why those events were necessary. In the same way there is no necessity for the exposure of urban communities to asbestos fibre. That is true of Canberra families living in homes with raw amosite secreted in the ceiling cavity for insulation; it is true for the employees of the National Library or for those school children meeting each morning in an assembly hall insulated with sprayed asbestos. The necessity will end with the vigilence of the public, of consumers, trade unions, parents and citizens who are unwilling to be the passive victims of the profit motive, and of indifferent governments and incompetent state authorities. It should also be remembered that the asbestos story could not have been possible but for an irresponsible medical profession whose indifference has only served to increase the wealth of greedy lawyers.

In its own defence the asbestos industry has always maintained that events have simply been beyond human control. Manufacturers were, so we are now told, captives of an imperfect technology, trapped by fallible medical knowledge and by the harsh realities of economic necessity. By its own judgment the industry has clean hands: after all, it must always deal with stupid employees who will not wear respirators or take sufficient care in using machinery. And yet a careful reading of the medical literature about asbestos-related diseases shows these claims to be groundless. There is no element of Greek tragedy about the asbestos story in Australia. But why those events took place and why the victims have experienced such difficulty in their struggle for justice provides an important lesson. The story of asbestos is an example of neglect, cynicism and the possibilities of abuse of labour where employees go unprotected by trade unions, government authorities or public scrutiny.

2

The Industry

The name *asbestos* is derived from the Greek word meaning unquenchable, but it is the mineral's ability to resist heat and cold which has made it so useful.

The name asbestos is applied to the fibrous forms of various mineral siliceous rocks. The most common of these is the serpentine group of which chrysotile or white asbestos is the most frequently mined. A second group known as the amphiboles contains crocidolite or blue asbestos and it is this form which was mined at Wittenoom in Western Australia. Most of the asbestos used commercially in Australia has been chrysotile, with amphiboles comprising both blue and brown asbestos, accounting for a little over 15 per cent of total consumption.

Asbestos has been used since prehistoric times and there is archaelogical evidence from Finland of pottery containing asbestos fibres dating from 2500 BC. Finnish peasants also used the material in insulating their huts. There are a number of references to the mineral from classic Roman sources, and in 456 BC Heroditus refers to asbestos as a cloth for retaining the ashes of the dead after cremation. We know that the ancient Greeks used asbestos in wick making, and that Charlemagne had a tablecloth woven from the fibre with which he used to entertain guests because of its fire-resistant properties. In the fifteenth century the fibre was used in French armour, probably because of its strength as a fabric, and it was first used in paper manufacture in Norway in the eighteenth century. Some of its more obscure uses include the weaving of asbestos thread to enforce virginity among African and Indian women through the sewing together of the labia, and the weaving of asbestos purses which Benjamin Franklin mentions in his diary from London. During the second half of the nineteenth century, asbestos was incorporated into the manufacture of Italian banknotes, and it came to be used widely in a number of European states in paper making. But the versatility of the fibre was not fully exploited until the revolution in manufacturing in western Europe and North America, which occurred at the end of the nineteenth century. The ever-widening diversity of products and, in particular, the growth in production of consumer durables made asbestos indispensable.

Asbestos fibre combines a number of qualities which together hold great appeal to industry. Its fibrous nature is combined with high tensile strength; it displays good adhesion and a resistance to attack by chemicals; it is impervious to heat and cold, to electricity and to abrasion; it is unappealing to vermin; and it resists distruction through vibration which makes it useful in any process involving packing or insulation. In 1876 a small firm was established in Scotland to manufacture asbestos yarn, but it was not until 1879 with the opening of the first mine at Thetford in Canada that there was large-scale exploitation of the mineral for commercial purposes. In its first year Thetford produced a mere 300 tons of fibre. A brief period of rapid expansion in the next five years was followed by a decline during the 1890s in the world-wide economic recession. In 1901 the huge American-based asbestos conglomerate, Johns-Manville, was founded. The firm's success helped to bring about the mechanization of mining and thereby the wide availability of asbestos products to consumers. In 1916 the British firm of Turner Brothers was established which, with Johns-Manville, was to dominate the industry for the next half-century.

By 1920 world production of fibre had reached 20,000 tons which is less than half of 1 per cent of output levels for 1976. Although mining and manufacture was centred initially in North America and the United Kingdom, South African mines began producing blue asbestos fibre in 1910. Output remained low until the early 1940s, when wartime demand for the material in gas mask production and the introduction of new mining techniques led to a rapid increase in production. In the first decades of the present century there was also some mining in Russia but, after the revolution, methods remained primitive and it was not until the 1960s that the Soviet asbestos industry became modernized. In Australia the first recorded production of asbestos fibre occurred at Jones Creek, a small town near Gundagai, in 1880, and there were also a number of small mines begun in Tasmania in 1889.[1] Over the next forty years various mines were established in South Australia, Western Australia and New South Wales, but none of these assumed commercial significance. Blue asbestos was mined in Western Australia and also at Robertstown in South Australia. The demand for fibre by Australian industry has followed much the same pattern as that found overseas, with almost 90 per cent of the fibre consumed being white asbestos. Most of the mineral has been incorporated into asbestos cement products although, on a small scale, there has been some weaving of asbestos fabrics and the manufacture of insulating materials. In the years from 1899 until 1944 Australian production of white asbestos was a mere 6,388 tons. In the period from 1915 until 1944, 1,736 tons of blue fibre were produced.[2] These figures indicate the modest size of the mining sector with imported fibre satisfying the needs of local manufacture.

Australia shares with South Africa and Zimbabwe the distinction of being one of the few countries in the world possessing commercially significant deposits of blue asbestos. The mineral occurs in the Hamersley Ranges in the northwest of Western Australia and, although the fibre displays some similarity to South African varieties, it tends to be longer. In the 1940s deposits were mined at four sites: at Yampire Gorge, Wittenoom, Dales Gorge and Marra Mamba. Of these, only Wittenoom, evolved into large-scale production, whereas mines at the other locations were worked by individual prospectors extracting small quantitites of fibre with primitive machinery.[3] During the war years, chrysotile or white asbestos was exploited commercially at Baryulgil in New South Wales and at Zeehan in Tasmania by Tasmanian Asbestos Pty Ltd. The Tasmanian mine, like the mine at Wittenoom, was worked as a subsidiary of CSR, whereas Asbestos Mines Pty Ltd, which operated the mine at Baryulgil, was a subsidiary of James Hardie. In 1944 only 5 per cent of local Australian needs for asbestos fibre was supplied from Australian mines, and imports of white asbestos from Canada and blue and brown asbestos from southern Africa predominated.

Despite the smallness of Australian output, there was considerable optimism during the 1940s that local mines could become important and obviate the dependence upon imported asbestos. According to a Department of Supply and Shipping document published in July 1945, it was believed that blue asbestos from Western Australia could make this country largely self-sufficient in this important material.[4] That, however, did not come to pass, and by 1976 Australian asbestos mines, both blue and white, had managed to produce only a little over one-third of a million tons of fibre during their thirty years of operation.[5] In that same year Australian mines accounted for less than 1 per cent of world asbestos production. There are no reliable figures on the consumption of asbestos by Australian industry over the past fifty years, but it is likely that local patterns of usage followed closely overseas trends. This means that asbestos cement sheeting would account for the vast majority of all fibre imported and produced locally. Imported fibre from Canada and South Africa was used mainly in bonded cement sheeting, with blue asbestos proving popular in curved cement products because of its high tensile strength and durability.

Asbestos is the carcinogen to which people living in the industrialized states have been mostly commonly exposed. The exposure of large numbers of people has been brought about by the sheer variety of the uses of asbestos fibre. There is no family in Australia which has not incorporated an asbestos-based or asbestos-containing product into its daily life. Asbestos is present in asbestos cement used in the sheeting of walls, in pipes and in moulded products. It is present in numerous friction materials such as brake linings, where asbestos can comprise from 10 to 90 per cent of the volume of brake shoes. Asbestos fibre is used commonly in vinyl floor materials and in textiles designed to resist heat, fire and cold. It can be found in fire curtains, in conveyor belts, in screens and wicks and in some insulating ropes. In the past, asbestos has been used in the filtering of beer, and it has commonly been incorporated into rubber and plastic products, in adhesives and cements and in paints and sealants. Asbestos insulating material was used widely in domestic dwellings before the advent of fibreglass and other synthetics. Some of the more exotic uses of asbestos includes its incorporation in cheese making, gas masks, paper, gloves, the filtering of fruit juices, domestic oven and ironing-board insulation, berets, aprons, mailbags, padding for pianos, the lining of soldiers' helmets, in carpets, table-mats and in domestic stoves. It is known that asbestos was used in Australia in the manufacture of children's dolls, and asbestos fibre was at one time a common ingredient in talcum powders where it found its way into condoms. For a period asbestos achieved a kind of vogue. In the 1950s the owners of the London Colosseum sought to attract patrons by advertising the woven asbestos curtain which shielded the audience from the threat of fire backstage. Patrons of the Colosseum were apparently no more anxious about the airborne fibre flacking from the curtain than were parents in innumerable Australian homes during the same period by the chipped asbestos stove mat found in so many kitchens.

The Australian Industry

Asbestos manufacture in Australia has been dominated by a single firm — James Hardie Asbestos. While other companies have been involved at various time in the manufacture of asbestos products, no other firm has achieved major industrial expansion on the basis of this one commodity. The history of asbestos in Australia is therefore, above all else, the history of a single firm. That firm has, over the past decade, the decade of the international asbestos scandal, become one of the most visible of Australian companies. The James Hardie organization has devoted considerable energy to create a new persona for itself, and it is not coincidental that the Hardie-sponsored "Life. Be in It." campaign and the asbestos scare should have occurred at the same time.

The company was founded at the end of the last century by Andrew Reid. In 1892, Reid formed a partnership with James

Hardie who, before his arrival in Australia, had been involved in the tanning industry in England. Over the next decade, Reid and Hardie acquired control of a number of agencies in Sydney covering commodities in various industries, and by 1903 they had begun to import French asbestos cement products.[6] Asbestos cement was a new material which had been invented only three years earlier, and James Hardie was one of the first companies to see its potential as a building material. From that time until the outbreak of the First World War, Hardie was the sole agent for asbestos cement in Australia, and after a slow start the material soon attracted a wide popular appeal. In 1916, Hardie acquired sixteen acres of land near Parramatta with the intention of establishing a factory to mass produce cheap building materials. In the following year, the company registered its new Fibrolite product, and within three years the firm's dominance of the market in New South Wales had been established. Most of the product was made at the Camellia plant near Parramatta but, in 1920, Hardie was able to expand its asbestos cement output with the establishment of the Riversdale plant in Perth. In 1926 yet another major factory was set up, this time at Brooklyn in Melbourne's western suburbs. During this period, the company also expanded into the New Zealand building materials market. The major reasons for the company's success at this time was through the securing of government contracts for sewerage piping in New South Wales, Victoria and New Zealand. It was these contracts more than anything else which allowed the rapid expansion to take place.

One of the peculiarities of the Hardie organization during its first thirty years is found in the pattern of recruitment. Before 1939, most employees came into the firm on the recommendation of people already employed by the company. To secure a job at Hardies it was necessary to have a personal introduction. James Hardie was run very much as a family business. The Reid family also sponsored the employment of many young men from the Burnside Orphans Homes at Parramatta, who were taken in on the understanding that they would be guaranteed employment for life. These young men were commonly referred to as the "Reid Boys", and they played an important part in the firm's history.

In 1942 Hardie established a plant to manufacture insulating materials, but it was in the immediate post-war period that the firm's era of greatest expansion took place. The post-war demand for building materials was seemingly limitless, and the New South Wales Department of Works and Housing even turned to importing large quantities of fibro-cement sheeting from Italy because local firms, notably James Hardie, were unable to keep pace with demand. Large public works programmes under the State Housing Commission, the Post-Master General's Department, the Metropolitan Water, Sewage and Draining Board in New South Wales and shire council contracts were the most lucrative parts of the company's market.[7] This expansion encouraged technical innovations and the firm was quick to introduce a wide range of new materials of which asbestos flux was but one. An indication of the extent of the Hardie success is suggested in a Tariff Board report of 1955 which contains the following comment: "For the three years ending June 1955, houses sheathed with asbestos cement sheets total more than 25% of houses built in Australia, and nearly 50% of those houses in New South Wales and Western Australia in that period."[8] James Hardie Asbestos dominated that market.

During its history, James Hardie has had a number of competitors, such as Wunderlich, with whom for a brief period it shared ownership of the Baryulgil mine. It had also to relinquish a small part of the asbestos market to the Australian giant CSR which, through a subsidiary, operated the blue asbestos mine at Wittenoom. The Colonial Sugar Refinery was founded in 1885 and the firm was based, as its name suggests, on the Queensland sugar industry. In 1936 CSR established a pilot plant for fabricating wallboards from bagasse, which is the crushed fibre residue formed in sugarcane processing. Although this venture had little success, the firm remained committed to entering the building materials industry, and in 1941 CSR established a plaster mill at Concord in Sydney. At that time CSR also investigated the prospect of starting a major asbestos mine at Wittenoom to provide fibre for its own plants. The company had some experience in the asbestos industry through the small white asbestos mine at Zeehan which was operated by a subsidiary, Tasmanian Asbestos

Pty Ltd. With the establishment of its own fabricating plant in Perth, CSR was able to gain a footing in the Western Australian asbestos cement market. But like Hardie's other competitors, CSR was never in a position to challenge Hardie's dominance of the local industry. James Hardie had the experience, expertise and, most important of all, the state and local government contracts to exclude competitors virtually at will.

Asbestos fibre would arrive at Hardie plants in jute bags where it was stacked by hand. When ready for use, the bags would be emptied, again by hand, on to a mixing-room floor. There it would be blended manually with pitchforks in a process similar to the mixing of dry cement.[9] The material was then forked into a mechanical blender. From there the fibre was blown through a duct system into what was called "the asbestos room". When the room was filled the doors would be opened and the fibre forked by hand into wheelbarrows. The fibre was then weighed and mixed with water and cement. This kind of process was carried out at Hardie factories throughout the 1940s, the 1950s and the 1960s, and there is evidence that the same basic procedure was still being followed in some factories until as late as 1975.[10] The manufacture of asbestos piping generated dust as did other processes, such as the fabrication of brake linings which usually contained a high proportion of asbestos. The men engaged in each of these processes were exposed to high levels of airborne fibre.

The Hardie company began a series of innovations in the 1960s with the use of asbestos substitutes, such as cellulose, in a range of flat sheet products. This involved the establishment of a paper mill at the Camellia plant in 1964, which made the product brand-named Hardiflex. Early research into asbestos substitution began before the international health scandal about asbestos and cancer, and it enabled the company in the 1980s to make a comparatively easy transition from reliance upon asbestos to the use of cellulose. The company was also assisted in this shift by the strength of its research and development departments. As early as 1934 Hardie established a research division at Camellia, and it was there that the first silica and later cellulose substitutes were developed. Even so, in monetary terms, the conversion costs for the company have been high and, at Camellia alone, the final outlay may have been as much as $3 million. James Hardie's success in surviving the decline of the asbestos industry is characteristic of the firm's adaptability and the political acumen of its management. One of the reasons for the firm's success from its early days has been its ability to control the market, and to remain immune from the imposition of import duties on the essential ingredient for its products, namely asbestos. The competitors for Hardie's share of the Australian market have, however, faced the daunting prospect of high tariff barriers. In the case of finished asbestos products, tariffs have been in existence from the date of Federation. From 1901 until 1928, the average tariff on all such products was 25 per cent. This fluctuated according to various categories, with member states of the British Empire paying between 12 and 37 per cent.[11] Such protection existed for the firm throughout the life of the Baryulgil mine, and it was not until as late as 1976 that the tariff was changed to a flat rate of 15 per cent. This tariff protection allowed Hardie to undersell all imported asbestos cement products on the local market, while guaranteeing the firm duty-free access to asbestos fibre. The one major threat to the company's control over the Australian market, Wunderlich, the original co-owner of the Baryulgil mine, was outmanoeuvred during the 1950s by a series of Hardie strategies and, subsequently, was never able to challenge Hardie's dominance of government contracts for asbestos cement products.

As early as 1947, Hardie sought to guarantee itself a supply of fibre by acquiring a share in a South African mine. In that year it purchased a substantial holding in Cape Asbestos Company Ltd, which then held leases covering blue and brown asbestos deposits in the Union. In October 1954, James Hardie bought all remaining shares in the company, thereby making it a wholly owned subsidiary. During the same period, Hardie purchased a holding in the Canadian company Cassair Asbestos Corporation of Canada. These manoeuvres were each part of a conservative strategy designed to guard against the prospect of a failure of supply of the company's most important raw material. There was, however, at this time no intention to internationalize Hardie industries, and the firm was content to remain an Australian company with control over the local and New Zealand markets.

The success of Hardie in retaining its market position can be gauged by its defence of a Tariff Board inquiry into imported asbestos fibre held in 1954. The inquiry took place at the request of CSR, which asked for the imposition of a 40 per cent protective tariff on all imported fibre. CSR's intention was to force Hardie, along with all local manufacturers, to rely upon fibre from CSR's own mine at Wittenoom. Under the terms of the scheme proposed by CSR, local manufacturers would have received remittance on the tariff duty if they used at least 15 per cent of Australian asbestos in their products. Hardie fought against the tariff, claiming that it would cost the company as much as £808,000 each year.[12] Even if it chose to rely upon CSR's Wittenoom fibre, the annual cost would, so Hardie claimed, have been in excess of £270,000. James Hardie won the case and no tariff was imposed. In retaliation, Hardie agreed to buy fibre from CSR but only after a substantial lowering of the price, and from that time on it used considerable quantities of Wittenoom asbestos. The CSR manoeuvre did, however, have one other consequence, for it encouraged Hardie to increase the capacity of the Baryulgil mine and, in 1958, a new mill at that site was opened at a cost of £70,000.[13]

Under the shelter of tariff barriers, government contracts and its dominance of the domestic market, James Hardie enjoyed a period of expansion during the decade from 1958. In 1959, it purchased Better Brakes Holdings Ltd, which had previously controlled distribution of asbestos brake shoes from the Camellia plant. In 1960, Hardie secured a series of lucrative contracts with the New South Wales government to supply friction materials for the states' railways. These particular contracts encouraged the firm to establish a separate friction materials division and, to this end, it formed a joint company with the British giant Turner & Newell. This joint company, Hardie Ferodo Ltd, was founded in April 1962. Eventually, Hardie acquired Turner & Newell's 40 per cent share, and the firm then became a wholly owned subsidiary. Running parallel with expansion and diversification within the domestic market, Hardie also made efforts to become an international firm. In 1964, Hardie entered into an agreement with a Malaysian, European and British consortium to establish an asbestos cement plant in Malaysia under the name United Asbestos Cement Berhad of Malaysia. The firm was launched as a public company in 1964 under the management of Hardie in joint ownership with the Malaysian government and Turner & Newall. Since that time Hardie has purchased the Turner & Newall share, thereby making the company in effect a Hardie subsidiary. In achieving this control, Hardie received substantial assistance in the form of location grants from the Department of Trade and Resources. Apparently, the department viewed the venture as being of some national importance.

There was further internationalization of the firm in 1972, when Hardie moved into the Indonesian market. The $10 million project involved an asbestos cement building materials and piping plant, in which Hardie now holds a 75 per cent share with the remaining interest being held by Indonesian shareholders. In combination with the plant in Kuala Lumpur, the Jakarta factory gives Hardie a monopoly in asbestos cement products for the expanding and lucrative South-East Asian market. It is expected that over the coming decade, this market will remain the only readily accessible outlet for asbestos-based products, as consumer resistance, trade union pressure and government legislation each takes its toll on the appeal of such commodities within Australia.

Throughout the 1970s, James Hardie continued to consolidate its hold on the local asbestos market. In 1977, it took over its major competitor in the asbestos cement building products field, Wunderlich. At the time of the takeover, Wunderlich controlled more than a quarter of the New South Wales market, and nearly one-third of the market in other states. This merger gave Hardie an effective monopoly in all states with the exception of Tasmania, where a local producer still retains a major share of the asbestos cement market. This degree of control gave Hardie the ability to dictate the price of asbestos cement products, which was reflected in a growth in profits during the 1970s. That period of growth ended, however, with the asbestos health scandal. The scandal resulted in the transformation of Hardie from a company based on asbestos products to a highly diversified corporation with interests in a myriad of enterprises ranging across numerous industries. At the dawn of this new phase in the company's history, James Hardie had established itself as one of the truly great national companies whose products, most of which contained asbestos, could be found in most Australian homes. John Reid, the company's chairman, made reference to that achievement in a public statement in 1978, when he said: "Every time you walk into an office building, a home, a factory: every time you put your foot on the brake, ride in a train, see a bulldozer at work. . . . Every time you see or do any of these things, the chances are that a product from the James Hardie Group of Companies has a part in it."[14] Given the presence of asbestos in most of the products to which he refers, John Reid's suggestion presents a chilling prospect. And yet, less than twelve months later, the company took the unusual step of changing its name from James Hardie Asbestos Ltd to James Hardie Industries. The change was believed necessary in order to distance the firm from its past as an asbestos-based enterprise at a time when asbestos had become recognized publicly as a health hazard. It was also a recognition of the intention that James Hardie should no longer rely upon asbestos as its staple.

In the years of its prosperity, the 1970s, James Hardie had some political experience of the new costs associated with asbestos manufacture. The firm's Parramatta factory incurred several fines under the Navigable Waters (Anti-Pollution) Regulations, for discharging asbestos-contaminated water into the Parramatta River. The fines were trivial but, to its credit, the company did respond to the ethos of the Clean Waters Act of NSW 1972 by installing expensive filtration systems to eradicate the problem at Camellia. By the end of the decade, James Hardie was facing for the first time the economic and political consequences of forty years of asbestos manufacture. In March 1979, Hardie reached agreement with the Federated Miscellaneous Workers Union to pay voluntarily a lump sum of $14,000 to all Hardie employees who had compensatable asbestos-related diseases: that is a condition which was recognized as such by the Dust Diseases Board of NSW. Over the next twelve months, other trade unions, including the Amalgamated Metal Workers Union (AMWU), reached similar agreements with the firm. These agreements were symptomatic of the changing climate in which James Hardie was now operating. But it was a climate which had changed almost imperceptibly in response to long-established medical knowledge about the effects of asbestos fibre on the lungs of workers and consumers alike.

By its own admission, James Hardie has been careful in following the growth of scientific evidence about asbestos as a carcinogen. When this question was raised at an international conference in New York in 1964, the company claims it took immediate action in introducing more stringent dust-control measures at all plants. It also established a medical surveillance scheme for its employees.[15] According to interhouse documents, James Hardie took its responsibilities on the issues of worker health and safety so seriously that, in March 1965, one of its senior officers commented, "Quite apart from any legal requirement, the Company accepts the moral obligation of ensuring that the health of its operators and staff is adequately safeguarded".[16] Apparently, the company was soon satisfied that it had done everything necessary to assuage its own highly developed sense of responsibility. The annual report for 1979 also shows that James Hardie had, to its own satisfaction, resolved all problems associated with asbestos and worker safety. The report contains the following comment: "Injury to health caused by asbestos dust is rare and only becomes apparent after many years. The cases now emerging had their origins many years ago before medical knowledge had reached its present levels, and the Company is taking proper and appropriate care of those cases."[17]

James Hardie now claims that in the 1960s it had established a sophisticated and well-run industrial hygiene department to ensure the safety of workers employed in asbestos cement manufacturer. Dust control was, so the firm now claims, a feature of all Hardie factories.[18] The company was also at this time in contact with Turner & Newall and Johns-Manville, the British and American conglomerates, with whom it exchanged technical information. Presumably, given the firm's concern with worker health and safety, that information included material on the vital subject of occupational health.[19] James Hardie's medical officers did keep abreast of the latest developments in the United States and the United Kingdom, and one of its senior personnel, a man who was intimately involved with the monitoring of worker helath at the Baryulgil mine, published original material on the subject in *The Medical Journal of Australia*. Writing in that journal in early 1974, Dr S. F. McCullagh argued against the possibility of any threat to health arising from the asbestos contained in car-brake linings. According to Dr McCullagh, asbestos has always been naturally present in the environment and in drinking water and therefore it cannot be considered harmful. Furthermore, any threat posed by the mineral must be placed in the context of the preventative measures taken by a concerned and responsible industry, of which Dr McCullagh was an employee. He then goes on to explain, "The industry is well aware of the hazards of asbestos, and having briefly reviewed these I think we should also remember that, if we considered no more than its fire-retardant properties and its use in brake linings, asbestos has saved more lives than it has claimed. With the great improvements in the standards of industrial hygiene over the past decade this credit balance, if I may so call it, will increasingly grow more favourable still."[20] Dr McCullagh's comments were made more than fourteen years after the dangers of mesothelioma were publicized internationally and some twenty years after the relationship between lung cancer and asbestos was documented in the United States, not for the first, but for the second time. Conclusive knowledge about asbestosis, a disease of the lung, preceded Dr McCullagh's article by more than fifty years, and it is only proper to judge his comments in the light of this pre-existing knowledge. It is also necessary to weigh his belief in the Australian asbestos industry's good faith by examining the extent to which industry had finally come under effective public regulation.

It was not until 1947 that safety codes in New South Wales under the Factories and Shops Act were extended to cover asbestos cement roofing manufacture. In that year there were a mere sixty-seven factory and shop inspectors, and only fourteen industrial and factory inspectors for the state as a whole. In

1947, those officers issued nearly nine thousand notices of complaint under the Act. Given the number of factories and the number of inspectors, there was little possibility for the effective regulation of many plants, including those within the James Hardie empire.[21] Despite this fact, a major inspection of dust levels at Hardie plants in New South Wales was carried out in 1957 by the Industrial Hygiene department on behalf of the Department of Public Health. That survey was intended to be industry-wide, but given Hardie's dominance of asbestos cement fabrication, it can be taken as a survey of Hardie plants alone. The report, for all its obvious deficiencies owing to time, lack of staff and expertise, is nonetheless valuable in giving a profile on the industry in New South Wales at the time. It also accords some insight into the hygiene practised at Hardie factories.

According to the report, the asbestos cement industry in New South Wales employed just under one thousand people processing asbestos fibre imported from South Africa, Western Australia and Canada into asbestos-based products.[22] It also refers to a small amount of fibre originating from the Hardie-owned mine at Baryulgil. The department found that, at Hardie factories, bags containing blue, brown and white asbestos were blended by "asbestos gangs" and were then shovelled by hand into pulverizers. The asbestos was then conveyed to hoppers. From there, the material was raked by hand into wheelbarrows and taken to the mixing area. The authors of the report comment that the men engaged in this part of the process were exposed to some airborne dust, whereas the later stages in the preparation of cement products was wet and therefore relatively dust-free. Dust readings taken at the bagging section measures 24.7 mp/cf. Most readings at other parts of the plant, however, were considerably lower and varied between 7.8 and 0.2 mp/cf.[23]

Officers from the Industrial Hygiene department also undertook a survey of 175 employees out of a state total of some 960 men working full-time in the Sydney area. Most of those men had been in the industry for less than five years. They had also been subject to constant movement from one job to another within individual plants. Therefore, the survey contains no index of exposure enabling the researchers to distinguish between

workers who had been subject to varying levels of dust. Predictably, because of the brevity of employment in the industry for
most of the workers surveyed, X-rays of all the subjects revealed
only two cases of asbestosis. The authors were surprised, however, to discover a large number of cases of other respiratory diseases, which were found in twenty-two subjects.[24] Half of these
involved chronic bronchitis, a disease which twenty years later
was to be found in epidemic proportions among workers at Hardie's Baryulgil mine.

The report's conclusions are largely optimistic and the authors
are satisfied that there is little chance, given the existing dust
levels, for workers to contract asbestosis. The one note of warning refers to specific sites within plants, where dust was found to
be high. The authors write, "Nevertheless, processes were found
where the dust concentrations were high enough to lead inevitably to the development of asbestosis, provided the employees
remain long enough in these particular jobs."[25] The report then
recommends various dust-control measures, such as the mechanization of dusty jobs, the introduction of water sprays and exhaust ventilation, and the regular medical surveillance of all vulnerable workers. Why the authors concentrated exclusively on
the question of asbestosis, and chose to ignore the threat of lung
cancer, already established in the 1947 British report by Merewether, is not explained. This omission is important because of
the known lower levels of exposure which can cause lung cancer.[26]
The reader can only assume that the Department of Public
Health was not competent and languished far behind orthodox
medical opinion in the United Kingdom and the United States.

The optimism of the Industrial Hygiene department is reflected in the workers' compensation insurance rates, to which the industry was subject at the time. Under the New South Wales
Workers Compensation (Silocosis) Act, employers were liable to
pay a fixed level of contributions for each £100 paid in wages for
the asbestos industry. In the period from 1947 until 1955, this
rate actually fell from £6 in 1945 to £4 in 1954.[27] Perhaps the fall
can be explained by the growth of the industry, or perhaps it
reflected the confidence of the public health authorities that
there was no hazard for such employees. Both explanations ap-

pear rather shallow in light of the growth of medical knowledge over the same period concerning the dangers from asbestos as a carcinogen. That knowledge also had little effect in discouraging the use of asbestos fibre as an insulating material. In 1959 the Director General of Public Health in New South Wales reported that his department had carried out dust-level tests in response to trade union concern about asbestos insulation. One of those tests was done while asbestos fibre was being sprayed into the ceiling cavity of a school assembly hall. If the material used was blue asbestos, the results from those tests suggests the presence of a real danger to the health of the children who, in successive years, were unfortunate enough to have sat in that hall.[28]

The voluntary compensation scheme announced by James Hardie in 1979 was syptomatic of the changed political environment in which the company has over the past eight years been forced to operate. At the annual general meeting in 1979, John Reid, the managing director of James Hardie, acknowledged that the firm was negotiating with the New South Wales Health Commission over the placing of warning labels on all asbestos products. He alluded to the controversy about the effects upon the health of Hardie employees, but he explained this was no cause for concern. The number of employees involved was small and, in Reid's opinion, over the preceding fifteen years only one hundred Hardie workers had contracted asbestosis. The context for Reid's remarks was to be found not in Australia where the issue had as yet to emerge but in events in the United Kingdom and the United States, where the survival of the asbestos industry was threatened by a flood of litigation which had first begun in the early 1970s. At the time of Reid's statement, more than 250 major American firms were subject to litigation with new claims being filled at a rate of over 100 each month. Reid's concern was that the problem would soon spread to Australia, and that James Hardie would thereby be faced with the same threat which was destroying its American counterparts.

The American market has always been dominated by a handful of conglomerates with Johns-Manville being the most predominant. In its annual report for 1980 the company acknowledged that it was currently the defendant, or the co-defendant, in more than five thousand asbestos health suits and that the number was increasing by some three hundred new cases each month. This figure had risen dramatically in the period since 1978 when new suits were being filed at a rate of only fifteen per week. Equally disturbing for the firm was the increasing amount being paid out in each successful claim. Prior to 1980, the average payout for a successful claim was $13,000, but by 1980 the figure had increased to $23,000. Despite its size, Johns-Manville was aware that it faced a difficult future as other smaller firms had been bankrupted by similar claims. In July 1982, UNR Industries of Chicago had been forced to take this step after a flood of litigation by more than seventeen thousand plaintiffs. Three months later, Johns-Manville followed the same path and in September filed for bankruptcy under federal law. At the time, the company had assets of more than $1 billion and its action was, as the management made clear, not due in any sense to a financial problem. Johns-Manville's president, John A. McKinney, later admitted that the move was intended to force the federal government to resolve an industry-wide problem in which many federal agencies, as consumers of asbestos products, were intimately involved. In this context Johns-Manville has argued that its action was taken in the best interests of shareholders, employees and even of the victims themselves.

Many smaller firms have experienced the same problems as Johns-Manville. The Keene Corporation's annual report for 1980 contains a thorough account of the firm's difficulties with asbestos disease litigation, which had beset the firm for almost a decade. By 1980 the costs of settling accounted for 3 per cent of annual sales and was expected to increase still further. Most of the six thousand outstanding claims related to the insulation of ships, and they included the full gamut of asbestos-related diseases: asbestosis, lung cancer and mesothelioma. Since 1975 such cases had cost Keene $5 million, and yet of the firm's annual sales, which are in excess of $200 million, only 2 per cent involved asbestos-based products. Keene's involvement in the asbestos industry arose through the activities of a subsidiary, Baldwin-Ehret-Hill Incorporated, which Keene acquired in 1968. This subsidiary neither mined nor manufactured asbestos products, but had bought and resold such products from a variety of producers, including the federal government. In response to the litigation, Keene followed two paths of action: it launched a series of lawsuits against its insurers to force clarification of their obligations for settling claims, and it also initiated a number of suits against those producers which had supplied BEH with asbestos products. In this second strategy, Keene claimed that the product manufacturers knew far more about the inherent dangers of asbestos than did BEH and that those companies had failed to warn BEH as to the likely consequences. Among the producers against which Keene took action was the United States government, which had supplied BEH with asbestos thermal insulation. The aim of this suit was for the recovery of costs already paid out by BEH in damages to former clients. The Keene case reproduces on a lesser scale many of the features of the Johns-Manville bankruptcy. It also tells a story of which Australian asbestos manufacturers have become so fearful.

Johns-Manville has denied that it had any knowledge of the dangers of asbestos until the publication in 1964 of Selikoff's seminal article on cancer among New York insulation workers. Following Selikoff's revelations, Johns-Manville, as it now claims, took immediate action to protect its workers from airborne fibre. It also claims to have taken similar action to protect the consumers of its products. Whatever the justification for the company's actions since that time, its legal and financial position is now largely dependent upon the status of the insurance policies which cover the firm for product liability and workers compensation. In 1947, Johns-Manville signed a series of insurance policies which provided a coverage of $360 million. This coverage represents the total amount of insurance available to the firm for the period from 1947 to 1976. Within this insurance, there is a mere $16 million allocated for primary coverage. Asbestos is currently classified under United States product liability law as "unavoidably unsafe" and such products must be accompanied by warning labels to the consumers, who can then decide if they wish to take the risk inherent in using such goods. The failure to provide a warning renders the product "unreasonably dangerous" and the liability for any resulting injury falls then upon the manufacturer. The manufacturer is assumed to have expert knowledge about the product and is therefore responsible at law for any injury. The consumer can only follow the advice of the producer. Most of the plaintiffs who have issued litigation against Johns-Manville were not employed by the firm but had used Johns-Manville products. The central issue in such litigation revolves around two questions: When did the hazard from the use of asbestos products become foreseeable, thereby making warning labelling necessary, and when did Johns-Manville communicate the extent of that danger to its consumers?

Many of the asbestos producers in the United States have followed much the same strategies in seeking to survive the asbestos scandal. Johns-Manville changed its name to the Manville Corporation, Keene Corporation became Bairnco Corporation and Manville's major competitor, Ray-Bestos, became the Raymark Corporation. Besides seeking to distance themselves in this way from bad publicity, producers have also tended to diversify out of reliance upon asbestos commodities. Some have established holding companies in order to erect a corporate veil between themselves and litigants. Most have sought to assign corporate strength away from the asbestos sector. Using this method, Manville has allocated less than one quarter of its assets to its asbestos subsidiaries, and the Chicago-based UNR corporation allocated only a small fraction of its assets to its asbestos unit, Unarco. Manville now claims that the parent company cannot be held liable as a defendant in asbestos litigation and, like Keene, Manville is suing its insurers for the settlement of outstanding claims.

The asbestos industry in the United States has made various proposals to the federal government in an attempt to protect itself against destruction. Manville's voluntary bankruptcy was designed specifically to force government intervention on the industry's behalf. The corporation wants workers to forego the right to issue litigation. Instead, it proposes the establishment of a trust fund to be administered by a board of experts, who would decide the merits of individual cases and set levels for compensation. The fund from which compensation would be allocated would be established from contributions by the industry and by the federal government. The industry's spokesmen have made much of the fact that the federal government was one of the major users of asbestos insulation products from which so many of the present claims have arisen. The use of asbestos insulation in "Liberty Ships" during World War II has proven the most fertile source of asbestos-related cancer. Both the Carter and Reagan administrations have sought to deflect such proposals as Manville's. The United States government is well aware of the huge financial burden that acceptance of responsibility would bring at a time of economic recession and fierce political debate over federal deficits.

Since 1979, the United States asbestos industry has been in decline, and annual consumption of fibre has fallen from 800,000 tonnes in 1973 to less than 300,000 in 1982. This decline is expected to continue, as more and more firms retreat into other industries. And yet Johns-Manville, or as it is now called, the Manville Corporation, has survived and is flourishing. By August 1983, barely one year after the bankruptcy was filed, Manville had recovered to a strong position. The company now holds a huge portfolio of cash and marketable securities and remains under court protection, which guards it from further litigation. Manville is also free from the payment of debts accrued before September 1982, and it has taken various steps to distance itself from the asbestos industry. Those steps have included the selling-off of the firm's Canadian asbestos divisions. In November 1983, Manville submitted to the court a company reorganization plan which allows for the establishment of a centralized claims office. This office would process all claims against the company. The strategic value of the proposal for Manville is that it enables the company to be effectively split in two: one sector would be liable for all asbestos litigation, while the other, comprising the bulk of Manville's holdings, would be free to carry on its business unhindered. After more than twelve months, no resolution had been reached on the Manville proposal. The Reagan administration has been careful to avoid accepting any liability for the industry's lingering crisis.

In the United Kingdom, product liability has not given rise to the same volume of litigation that is found in the United States, although in one case Turner & Newall, the British giant, incurred a $650,000 payout awarded by an American court. Despite having escaped the worst legal action, the British industry has, like its American counterpart, suffered a sharp decline. In 1982 Turner & Newall showed a pre-tax loss of £4.5 million, compared with a profit of £8.5 million in 1981.[29] Apart from the asbestos producers themselves, one of the major casualities in the United Kingdom could be the London-based insurance firm of Lloyds, which is believed to hold product liability policies totalling several hundreds of millions of dollars. Lloyds is primarily a re-insurer for other firms who have in the past sought to spread the risk associated with asbestos insurance. Even if many of the policies involved are held eventually to be invalid, Lloyds is still faced with the prospect of massive legal fees in defending itself against liability.

There are various reasons for the comparatively lower incidence of litigation in Britain compared with the American experience. In general, the social costs associated with the British asbestos industry have been less severe than for American workers, consumers and governments. And yet, in both environments, the asbestos industry is being confronted by payment for the hidden costs associated with more than forty years of successful trading in asbestos-based products. The sums awarded in compensation may appear dramatic in absolute terms, but they should not be viewed in abstraction from the total economic activity, including the corporate profits of the firms involved over that extended period. When seen in this way, the present rash of litigation can be given a proper perspective. Companies such as Manville and Keene have been willing to approach the United States government for protection, but they are not willing to refer to the wider context in which the asbestos scandal belongs: the context of successful corporate activity, and the constant expansion of profits over a period of several decades. For firms such as Turner & Newall, Keene, and Manville, the asbestos industry was highly profitable. But, in each case, those profits were maintained at the hidden social cost which the firms now deny is their responsibility.

The costs associated with litigation, even if including the legal fees involved, grossly underestimate the final costs generated by the asbestos industry. Many of the victims of cancer and asbestosis have never been diagnosed but, nevertheless, spent years in chronic ill health, lingering on in hospital wards, or struggling through limited lives on social security payments. In the United States, as in Australia, the burden of such payments has become a major political issue, because of the pressure they place on federal and state budget deficits, taxes and inflation. In the United States, the number of permanently disabled persons, as defined by the Social Security Administration, rose from twenty million in 1970 to over forty million a decade later. More than half of these people are supported at public expense through a variety of social security and social welfare payments. In 1981, the United States government paid out $17 billion to the disabled and their relatives. Some of these people are the victims of the industrial process and of an occupational hygiene malpractice. As the latency period for asbestos-related diseases is reached for those exposed after World War II, the number of people seeking medical care will rise. The costs for treating these victims is borne largely by the taxpayer, shifting once against the burden of payment from the corporate sphere of Johns-Manville and Turner & Newall on to the shoulders of an anonymous public. Any proper accounting of the costs of the asbestos industry must take full account of such medical care and social welfare subsidies. Those subsidies are in effect being made on behalf of the industry and thereby represent a form of relief.

It would be wrong to assume that the asbestos scandal in the United States has destroyed the industry overnight. Asbestos still remains one of the more important manufacturing specializations in the United States, West Germany, Japan and a host of other countries. Canada, the source of most of the world's chrysotile, still boasts the world's largest, single, open-cut quarry at the town of Asbestos in Quebec. In 1981, a major redevelopment plan was begun at Asbestos to expand the existing quarry's output and productivity. The industry will remain important to the Canadian economy, although in future the market for Canadian asbestos will shift away from the United States to Third World countries and to Japan. South Africa is still the world's third largest producer and the dominant supplier of blue and brown fibres. The three mining houses which control production in the Republic are General Mining and Finance Corporations, Cape Asbestos/Transvaal Consoldiated and, finally, Asbestos Investments Pty Ltd. There are also several smaller companies, but none has the ability of the larger firms to influence the market. South Africa refuses to publish official estimates of its asbestos exports, either by volume or in regard to the countries of destination. Virtually all local fibre is exported with the majority believed going to Japan, Spain, West Germany, Italy, Thailand and, until recently, the United States. Japanese consumers and trade unions have shown little interest in the carcinogenic properties of asbestos and there has been an increase in consumption during the past decade. Among the world's other producers of fibre are Zimbabwe, where blue asbestos is mined by a subsidiary of Turner & Newall, and the People's Republic of China, where annual production, centred in the Szechuan region, has recently been expanded.

The world's major producer of asbestos is currently the Soviet Union. Output and local consumption of the material have been increasing rapidly against the trend apparent in virtually all OECD states. With the expected modernization programme in the mining sector, Soviet production is certain to increase over the coming decade. In 1979, the huge Kiyembay project commenced operations. The project is being financed by a consortium of six COMECON nations which, when the mine is in full operation, will share some 200,000 tonnes of fibre each year. It is believed that Eastern European bloc countries have provided some of the labour as well as technical advice at Kiyembay. The remainder of the fibre will be exported to India and to Yugoslavia. In Russia there is no environmental or occupational health lobby to constrain the use of asbestos materials. The Soviet public does not have access to information either about the presence of asbestos in consumer products, the conditions under which those products are made, or the existing medical knowledge of the hazards associated with all forms of asbestos. Soviet scientists, at least in official publications, tend to decry Western evidence against asbestos, which they view as groundless and exaggerated. There is no reason to expect opposition to the growth of the industry, both for the export of fibre and for local manufacture. Fortunately, in the West asbestos has not fared so well.

With the exception of Japan, use of asbestos in all industrialized states is expected to fall in the coming decade. Recommendations from the European Economic Community will see the disappearance of the mineral and its gradual replacement by substitutes such as silica and cellulose. Sweden and Denmark have already banned the use of asbestos, and the introduction of even lower levels of fibre in new asbestos cement products have accompanied the legal battles in American courts. Under the Carter administration, new regulations combined with bad publicity to reduce the appeal of all asbestos products to consumers. The process has been slowed down to some extent by the less exacting attitude adopted under the presidency of Ronald Reagan. Even so, the industry has little chance for recovery. Effective campaigning by producers of substitutes, such as fibreglass and ceramic and silica fibres have further eroded the market previously enjoyed by asbestos producers, with glass fibres making the greatest impact. It is expected that, by the turn of the century, total consumption of asbestos in the United States will fall below a quarter of a million tonnes per annum.

Despite its misfortune, the Amercian industry has not stood idly by in the face of criticism from its opponents. In 1970, the Asbestos Information Association of North America was founded. The association is a non-profit publicity and promotions body which represents more than fifty American and Canadian firms. Most of the association's energies have been spent on lobbying support and in trying to counter growing public awareness of the dangers of asbestos products. But, like the Asbestos International Association, which has representatives in twenty-six countries, the tide of criticism has been simply overwhelming. In all OECD states, the industry now faces serious obstacles which have all but destroyed the credibility of its products. Asbestos now has a negative image among consumers. This image has been bolstered by activities of trade unions concerned about the welfare of their members forced to work with asbestos materials. Extensive litigation arising from the past and the high costs of complying with more stringent governmental regulations have eroded much of the profitability previously enjoyed by the industry. There is now little research and development of new products and the market for asbestos is shrinking under the weight of more attractive substitutes. The makers of those substitutes have added their voice to the criticism, thereby hastening the industry's decline. In combination, these factors herald the end of the asbestos market in most Western states, and they have encouraged the industry to seek new markets among the less discriminating consumers, who are to be found in the developing world.

In many Third World countries, there is neither the government apparatus nor the political will to seek to regulate the activities of international firms, which bring with them the promise of investment and jobs. In countries such as India, Thailand, Taiwan and Mexico, governments are enthusiastic about the establishment of new plants and appear to have less concern than their counterparts in the West as to how those plants are managed. They also care less about the possible ill-effects from the goods produced. A telling example of this lack of concern is found in the tragedy at Bhopal in India, which involved an American multinational enterprise and revealed the feeble regulation of that company's activities by state authorities. Asbestos cement piping has great appeal in Third World states. Such products are cheap to produce, they are durable and the technology involved in their manufacture is less complex than that associated with substitutes. For Western firms, the costs of compliance with local occupational health and environmental standards in countries such as India or Indonesia are far lower than in the metropole. The new states therefore present an attractive location for production and a readily available and expanding market for the product itself.[30] In the next twenty years, it is predicted that asbestos-based products will find a welcome market in the Gulf states, Taiwan, South Korea, the Philippines, the Caribbean and in the ASEAN nations. This shift to the Third World is so attractive that American and British firms are not the only ones to have noticed the potential to be found away from traditional markets.

In Australia, as in the United States, the next decade will see a steady decline in the production and use of all asbestos products. The James Hardie organization has accepted that the new technologies for handling, using and disposing of asbestos will not be sufficient to counter the declining appeal of the fibre. In August 1980, James Hardie announced the sale of its share in a Canadian chrysotile supplier, Cassair Resources. The share was sold for over $8 million and is indicative of the firm's intention to move away completely from the asbestos industry. During the same period, Hardie undertook an extensive retooling of its plants in substituting cellulose for asbestos in the manufacture of cement-sheeting products. The first of these so-called SX products were marketed in May 1981 from the company's Perth plant. In 1982 alone James Hardie spent more than $5 million in this conversion process. Within the space of a mere two years an organization which had so recently displayed the word "asbestos" as part of its trade name had virtually left the asbestos industry. The only products still made by Hardie which contain asbestos are a small range of high-pressure cement pipes. Asbestos is essential in these particular pipes, because of its high tensile strength and it will be some time before cellulose or any other substitute is suitable for its replacement. In 1984, less than 10,000 tonnes of asbestos fibre were used in Australia, which is a sharp decline from the 70,000 tonnes consumed in 1980.

The local industry's publicity and promotions organization, the South Pacific Asbestos Society, closed its doors on 31 December 1984. SPAS, which was sponsored by the local manufacturers, had been active over the past eight years in all public inquiries relating to asbestos. Like its American counterparts, SPAS sought to deflect criticism of asbestos and to present the virtues of the industry and its wares to hostile consumers and government departments. The closure of SPAS means that the Australian industry no longer has a permanent secretariat. James Hardie, which had provided the bulk of the funds for the society, retains an interest in asbestos through its subsidiaries in Jakarta and in Malaysia, which it established in conformity with the practice of British and American firms. James Hardie has been adamant that industrial hygiene at both of these off-shore enterprises is of the highest standard and compares favourably with Australian requirements. Whatever the truth of such claims, they are unlikely to be tested by industrial hygiene officers in Indonesia or Malaysia, both of which are Third World states.

The asbestos story in Australia does not end with the withdrawal of James Hardie from domestic production. The immediate victims of the industry's past forty years are still among us: the men and women who worked in asbestos cement factories, the miners and their wives and sometimes even their children, who lived in environments saturated by asbestos fibre; the brake repairmen, the railway workers, the naval dockyard employees, the home handymen and the men and women in the construction industry. In the next generation, the victims of asbestos-related diseases will be drawn from among the ranks of all these people. They may also come from those who work in offices and live in homes in which asbestos insulation has been sprayed into ceiling cavities. Asbestos insulation is known to be present in such notable buildings as the National Library and Parliament House in Canberra, and the Treasury Offices and BHP House in Melbourne. The circle of disease caused by the mineral with so many uses is ever widening as is the political stage on which the central issues of private advantage and public cost arising from asbestos must eventually be resolved.

3

The Medical History

Awareness of the dangers posed by the inhalation of asbestos fibre existed long before the emergence of medical evidence establishing conclusively the nature of the hazard.[1] In the literature from the ancient world there are numerous references to illness caused by the inhalation of various dusts. In classical Greece no distinction was made between the effects on the lungs of different substances, and all pulmonary diseases arising from dust were referred to as phthisis. In the sixteenth century European physicians were interested in the effect of "pestilential air" and numerous papers were published on disease among miners, smelters and grain workers who were known to suffer from particular disorders of the lung. In 1597 the first surviving study of occupational health among miners was published in Switzerland, but it was not until the nineteenth century that sustained research was carried out into the health of miners.

In the 1830s and 1840s various studies were published in Britain on the effects of dusty trades on longevity. Those studies included reference to the work of stonecutting, which was known traditionally to be a hazardous trade. In 1861 a British Royal Commission was established to inquire into the health of British miners following years of debate about the ill-effects of coal and other dusts. This commission in its final report reflected ruling medical opinion which presumed that illness among miners was due principally to bad hygiene, rather than to the action of any element in the dust itself. Subsequent research into morbidity among rock-drillers led the way to a recognition of the effect of dust, and in 1870 the term "silicosis" was coined to describe the condition arising from exposure to silica. In Britain, as elsewhere in Europe, coal and silica were the most commonly encountered dusts in the mining industry and, therefore, most illness was attributable to those particular substances. Silica was also identified early because of the scarring or fibrosis it induces, which leads to a stiffening of the lung tissue. Recognition of the dangers posed by asbestos dust followed soon after.

The earliest British government document, which mentions the hazards of asbestos fibre, is a factory inspector's report of 1898. In that report the inspector comments, "The evil effects of asbestos dust have also attracted attention."[2] Four years later in a survey of dangerous occupations published in New York, reference is made to asbestos as being one of those substances known since antiquity as being injurious.[3] Such scattered references as these do not in any sense constitute a ruling orthodoxy about the hazards of asbestos. The references in works from antiquity until the third decade of the present century are no more than half-recognized suspicions, which may or may not have had wide currency among physicians. Until the 1920s the asbestos industry was so small that no individual physician would have come across a sufficient number of cases of asbestosis or lung cancer among asbestos workers to be alerted to the connection. The industry's most strident critics have presumed to find in these scattered statements proof that asbestos was already established as a dangerous substance by the turn of the century. Those critics have not, however, been able to provide anything more substantial than a few meagre citations, which together constitute nothing more than a suspicion.

The breakthrough in identifying the effects of asbestos began in Britain in 1906 with the work of Montague Murray. In evidence before a British Departmental Commission on Industrial Disease, Murray described the case of an asbestos worker who had been in the industry for ten years. Murray's patient was the sole survivor of a group of ten workers employed in a carding room, all of whom had presumably died of asbestosis. A post mortem of Murray's patient revealed pulmonary fibrosis, and his is the first documented case of a death resulting specifically from asbestosis. A survey of British industry in 1910, which included a number of asbestos factories, failed to discover any specific danger, and it was not until 1924 that a second death from asbestosis was identified.[4] In publishing this case, W. E. Cooke commented that physicians treating asbestos workers had long suspected a connection with chronic bronchitis and fibrosis, and that it is known to cause the later disease in guinea pigs. The case cited by Cooke was that of a young woman, aged 33, who had been employed in an asbestos factory. A clinical examination showed that her fibrosis was extensive. Over the next five years the *British Medical Journal* contained a number of articles on the subject of asbestos and pulmonary disorders.[5] These works suggest that asbestos-related disease of the lungs is quite distinct from that present in patients suffering from tuberculosis. This spate of articles was indicative of a growing concern among physicians.

In 1930 the first major study of the effects of asbestos on occupational health was published in Britain.[6] The study was carried out by Merewether and Price and was initiated by the Factories Department of the Home Office after the discovery of fibrosis in the lungs of an asbestos worker named Seiler. Seiler's was only the third such case following those already identified by Montague Murray and Cooke. The Factories department wanted to know if Seiler's death was exceptional, or if it formed part of an underlying pattern of disease among workers in the industry. Merewether and Price confined their investigations to workers exposed to pure asbestos fibre and excluded subjects who were known to have come into contact with other fibres such as cotton. They discovered that perhaps as few as 2,200 workers in British factories fell within the scope of such a study and from these they chose a sample of 352 subjects. In each case the authors sought to include those who had the longest period of employment in the industry, and yet 60 per cent of the sample was drawn from workers with less than ten years exposure. The authors explain that, even so, the sample is representative, given the age structure of asbestos workers and high turnover characteristic of the industry as a whole.[7]

Each subject was given a full medical examination, which included chest X-rays and the taking of a complete medical history from childhood onwards. Merewether and Price found that of the 374 patients, 105 had diffuse fibrosis.[8] After making adjustments to the sample to exclude other possible causes, they concluded that over one-quarter of the workers had contracted the disease because of their occupation.[9] After five years of employment the incidence of the disease appeared to rise sharply and after a period of ten years its rise was in " . . . almost geometric progression".[10] The authors conclude that the inhalation of asbestos will, over a period of years, result in a serious fibrosis in those air-cells of the lungs where the asbestos comes to rest. In the authors' view the disease displayed an insidious onset with the normal reserve capacity of the lungs masking the effect for some years. Over time, the victim will have a progressive loss of breath in a disease which is distinguished from silicosis by its mode of distribution in the lungs and by its more rapid development. They have no doubt that the disease can be fatal.[11] The authors could discern, however, no variation in the disease when found in association with the different types of asbestos fibre used by British industry. But they did discover that the disease was dose-related and that for workers exposed to high concentrations of dust there was a greater chance of illness.[12] However, to Merewether and Price this very property of asbestos suggested the means for future prevention.

Merewether and Price concluded that if the dust levels found in the industry were reduced then the length of time before the appearance of disease would be lengthened. Therefore, the perfection of the means of dust control could lead to the elimination of risk in the industry. In 1930 there were 160 factories in the United Kingdom manufacturing asbestos products. Of these, Merewether and Price found only 18 involved in weaving textiles from fibre. Most of the plants were small and employed fewer than five hundred workers each and there was little effort made towards proper industrial hygiene. The most dusty plants were those making mattresses using asbestos as a filler and as a covering material. This work was carried out almost entirely by hand with the asbestos cloth being cut into shape on open benches and the filling material being taken manually from open bins. When stuffed with filler the mattresses were beaten by hand or with a flat wooden bat. Almost every single operation in the productive process created dust.[13] Using mattress production as a starting point, Merewether and Price constructed a list of preventive measures which they believed should be taken in any plant using asbestos. They commented that the protection afforded by respirators was unsatisfactory as such devices gave rise to a false sense of security. Respirators were also uncomfortable, and because they prevent easy communication, which is so important in any work environment, employees cannot be relied upon to wear such equipment for any length of time. A better remedy was to be found in the use of efficient exhaust ventilation and the separation of those processes which generate dust from other sections of the workplace. The authors also recommended the use of sacks woven from closed material, rather than hessian bags which readily allow the escape of dust from the loose weave. The mechanical shaking of bags was also suggested since the process exposes workers to high levels of dust if done manually.

The recommendations in the report were made on the understanding that the dangers found in the asbestos industry were greater than those present in other parts of the manufacturing sector where dust is also a problem. On this subject the authors commented, "The asbestos manufacturers are clearly confronted with the necessity of attaining conditions in their industry which will ensure much less dust in the atmosphere than can safely be tolerated in many comparable trades not using asbestos."[14] The warning to industry could not have been more clear, and one can assume that Australian manufacturers of asbestos products would have been aware of the change in British industrial hygiene philosophy and policy which the Merewether and Price study established. That change in outlook occurred more than thirteen years before the Australian asbestos mines at Baryulgil and Wittenoom were opened.

Within eighteen months of the release of the Merewether and Price study, the British Home Office released a memorandum to all manufacturers on the subject of silicosis and asbestosis.[15] The document defines asbestosis as an industrial disease which is distinct from silicosis in its pathological and clinical appearance. It also describes in some detail its symptoms. Most important of all, the memorandum identifies the suppression of dust as the best means for the prevention of disease. It recommends the substitution of wet for dry methods of processing, the enclosure of dust-producing machinery and the substitution of mechanized for manual methods wherever possible. The document also suggests the importance of periodic medical examinations for all employees. The memorandum was widely circulated and, presumably, its contents were also brought to the attention of asbestos manufacturers in Australia, including the James Hardie organization.

Perhaps the best means by which to judge the level of medical and industrial knowledge in Australia during the 1930s and 1940s about the threat posed by asbestos is to examine the basic medical texts which were in vogue during the period. The best known of these texts is Lanza's *Silicosis and Asbestosis,*[16] which was first published in 1938 and remained in print for the next twenty years. In the preface to the book, Lanza is described among other things as an adviser on industrial hygiene to the Commonwealth of Australia. He was also an assistant medical director of the Metropolitan Life Insurance Company. Lanza was accepted internationally as a leading expert on the subject of pneumoconiosis, and presumably his work was as well known in Australia as it was in London, Paris and New York. In his book Lanza describes in some detail the aetiology and symptoms associated with asbestosis which he acknowledges as having first been defined by Merewether and Price. Asbestosis patients invariably display a persistent cough with little or no expectoration, but their most striking symptom is dyspnoea or shortness of breath. The clubbing of fingers and toes and anorexia are present only in the latter stages of the disease. In many patients, breathing becomes progressively more difficult and cardiac enlargement is common. Lanza points out that mis-diagnosis in cases of asbestosis is frequent with physicans often mistaking the disease for any one of a number of cardiac disorders.[17] Despite the date of its publication, *Silicosis and Asbestosis* also contains a warning that there may be a close relationship between asbestos and cancer of the lung, but Lanza is hesitant to draw any firm conclusions without statistical evidence.[18] At the time of publication

Lanza was aware of only seventy-eight recorded cases of asbestosis deaths in the whole of the United Kingdom in the period from 1930 to 1936. This paucity of statistics he admits provides only the most fragile basis for drawing any firm conclusions.[19] Even so, Lanza's work carries a clear warning for employers about rendering safe any work environment in which asbestos fibre is present. Most important of all, he explains that detectable changes in the X-rays of asbestosis victims usually precede the symptoms of breathlessness and reduced lung function. Clearly, this warning placed the responsibility on employers to periodically screen workers as a safeguard against an insidious disease.[20] Lanza, like his contemporaries, believed that the harmful properties of asbestos were probably due to the chemical composition of the material, and it was not until many years later that this view was finally discarded. Today it is understood that it is the morphology or structure of asbestos fibre which explains its pernicious effect on human tissue.

Recent research has established conclusively that the structure of asbestos fibres is the factor which allows the mineral to reach vulnerable parts of the lung. The morphology of asbestos enables individual fibres to penetrate more deeply than other airborne particles. Chrysotile or white asbestos fibres are soft, curly and pliable and tend to fragment into sub-fibres and fibrils. Crocidolite or blue asbestos fibres tend to be straighter and more rigid, and retain their consistency at even the smallest diameters. This allows the fibres to penetrate deeply into the lung, whereas white asbestos fibres will be trapped at higher levels and expectorated. If fibres are swallowed, some will be absorbed through the gastrointestinal tract, thereby entering the bloodstream. This enables the material to travel to any part of the body, and autopsies of asbestos workers have revealed the presence of fibres in the brain, the liver, pancreas, prostate, spleen, kidneys and thyroid glands. Asbestos is made more dangerous because of its durability in the body, rather than because of its chemical action directly upon tissue. Animal studies have shown that all types of asbestos are capable of inducing cancer, but the behaviour of asbestos upon human subjects is little understood.

Like silicosis, asbestosis is classified medically as a type of pneumoconiosis or dust-induced disease of the lungs. Once a foreign material enters the body, defensive cells gather to the site, thereby setting up an inflammation. If the irritation is prolonged a fibrosis or scar tissue may form. Such tissue is inelastic and over time will tend to shrink. In the lungs this kind of damage leads to a reduced function which may remain unnoticed for years because of that organ's excess capacity. If exposure continues and the scar tissue widens, the person will gradually become aware of breathlessness, and exhaustion will occur even after the most casual exercise. As the disease progresses the patient will become prone to other infections and illness, such as bronchitis and pneumonia. The scarring of the lungs creates a resistance to the normal flow of blood, requiring the heart to work harder pushing blood through the lungs. A totally incapacitated asbestosis patient will spend his or her days in bed breathing oxygen-enriched air through a face mask. The patient will be unable to do any of those things which healthy people take for granted, such as walking, gardening, or playing with children. The immediate cause of death will usually be from respiratory or heart failure brought on by a secondary infection.

Asbestosis is but one of a number of diseases caused by exposure to asbestos fibre. It is now accepted that all forms of asbestos are carcinogenic in animals. Tests carried out with mice, hamsters, rabbits and rats have established conclusively that asbestos provokes tumours.[21] In the case of rats various types of asbestos produce cancer of the lung, liver and mammary glands following not only inhalation, but also after swallowing. In the case of human subjects asbestos is believed to present a hazard not only for cancer of the lung but also for cancer in various other organs of the body including the larynx. Asbestos is such an effective carcinogen that an IARC monograph published in 1976 contains the following comment: "At present, it is not possible to assess if there is a level of exposure in humans below which an increased risk of cancer would not occur."[22] Subsequent research has reinforced this view.

The growth of knowledge connecting exposure to airborne asbestos with cancer in humans has been tortuous. There is colloquial evidence from antiquity about the high incidence of what would now be termed bronchiogenic carcinoma among slaves employed weaving asbestos fabric. But it was not until the third decade of the present century that the true hazards of the material began to be unravelled. There was a series of articles published in British medical journals in the 1930s which drew attention to the co-existence of asbestosis and lung cancer, but these were scattered, and no firm conclusions were drawn by their authors. The first definite evidence of a causal link is found in a 1947 British report from the Chief Inspector of Factories.[23] This study was carried out by Merewether and was based upon an analysis of 235 autopsy reports of subjects whose acknowledged cause of death was asbestosis. Those deaths occurred in the period from 1924 until 1946. Merewether found that cancer of the lung or pleura was present in 31 cases, that is, in 13.2 per cent of the 235 subjects. There was a higher incidence among male subjects, of whom 17 per cent suffered from cancer. These figures become significant when compared with the incidence of cancer among silicosis sufferers. In the period from 1930 to 1946 Merewether found that cancer of the lung was present in only 1.32 per cent of deaths attributed to silicosis. Clearly there appeared to be a correlation between exposure to asbestos and cancer of the lung. Merewether's report was published in 1949 and it was soon followed by a series of academic articles which substantiated his findings. Those studies culminated in the work of Richard Doll in 1955.[24] Doll's study was based on analysis of necropsies, dating back to 1935, and he concluded that cancer of the lung was a specific hazard for asbestos workers.

In its defence the asbestos industry has argued that there was no general acceptance of Doll's findings until as late as 1964 when his work was validated by Selikoff's study of asbestos insulation workers. It is difficult, however, to reconcile the lack of precautions taken by the industry with the evidence contained in Merewether's much earlier report. Obviously the industry in the United States, Britain and Australia believed that it was under no obligation to lead the way in matters of industrial hygiene and that the only necessity for such action lay in governmental directives. In judging the industry's attitude, it is important to remember that the lapse in time between Merewether and Doll's publications was the period during which the industry expanded most rapidly in Britain and in North America. It was, therefore, the period during which more and more employees were exposed to asbestos and thereby to the risk of asbestosis and cancer.

Among the general population, asbestosis is a rare cause of death and not all workers who are exposed to even high levels of fibre eventually die from asbestos-related disease. Results from Canadian and American studies suggest that as few as 10 per cent of asbestos workers eventually will die from asbestosis, but the authors in each case admit there are various reasons such results may underestimate the true figures.[25] During the 1960s epidemiological evidence about the dangers of asbestos was scattered and, in most cases, studies were based upon only a small number of subjects. As Merewether and Price had commented thirty years earlier, the latency period between the contraction of the disease and its identification makes it difficult to carry out research. Even in the 1960s the latency period for asbestos cases arising from the industry's post-war expansion had not been reached and, therefore, insufficient cases had emerged to end doubt about the dangers of the material. Despite these obstacles, the connection between asbestos and cancer and, in particular, the dangers faced by asbestos workers who smoke were established conclusively in Selikoff's research into insulation workers in New York. Selikoff found that such people have a 92-fold greater risk of dying from lung cancer than non-smokers who do not work in the industry. The way in which cigarettes and asbestos work in tandem is not properly understood: fibre and smoke may act independently, thereby giving rise to an additive effect, or the asbestos and smoke may act synergistically creating a multiplicative effect. It is possible that asbestos fibre and cigarette smoke could react chemically, one with the other, to increase the likelihood of tumours, or the carcinogens in the smoke may sensitize lung tissue, thereby making it more receptive to the influence of fibre. Whatever the mechanisms involved, the consequences for asbestos workers who smoke have been well understood since 1964.

The facility of asbestos to cause cancer in humans is not confined to the lungs, and there is a growing body of research which connects asbestos exposure with cancer in other organs. There is now evidence that exposure increases the likelihood of cancer to other parts of the respiratory tract as well as to the stomach, colon, and rectum. Asbestos is implicated in cancer of the lip, tongue, salivary glands, mouth, small intestine, diaphragm, *rete testis* and ovaries.[26] There is evidence from Japan which suggests that rice consumption and cancer of the stomach are related. It is common practice in that country to add talcum powder during rice processing. Such powder invariably contains some asbestos, which would also help to explain the presence of chrysotile fibre found in the biopsy of gastric ulcers taken from Japanese patients who had never worked in the industry.

Because it is not possible in animal studies to control accurately the length and diameter of fibre, it is not known what particular type of fibre is the most damaging biologically. Blue asbestos is reputed to have a greater effect upon lung capacity and these fibres tend to be smaller and straighter than other varieties. There is also evidence from animal experiments that asbestos is capable of crossing the placental barrier in mammals. Asbestos is a common contaminant of talcum used as a dusting powder in condoms. This provides an obvious means for fibres to enter the uterus.[27] As yet no conclusive research has been carried out in this area.

The industry is correct in making much of the distinction between the presence of asbestos in the general and work environments and the availability of asbestos for exposure. In volume, most of the asbestos with which consumers come in contact is bonded into another material, such as cement or vinyl, and there is no immediate risk of exposure to airborne fibres. This distinction, however, does not guarantee that such products are safe. Asbestos, bonded into modelling clay, may present a risk if a child should place the clay in its mouth,[28] and all surfaces which contain asbestos are likely over time to be subject to wear. Consumer and environmental groups in the United States have focused their attention on the presence of airborne fibre in public buildings, such as schools where asbestos was used commonly during the 1950s and 1960s. In one such school, asbestos flacking from roof insulation was taken up by the ventilation system which circulated the material throughout the school buildings. An inspection discovered that the fibre was so thick as to clog the air filters. Samplings found levels of contamination of up to 3.8 fibre per ml,[29] a level far in excess of that which is now considered safe in EEC countries. Fibre from any number of products containing asbestos can be released into the air if the restraining sealant is broken through age, cleaning or friction. This is true of vinyl asbestos flooring, electric irons and toasters, domestic stoves, millboards, caulks, ironing-boards and hair dryers. Even if fibre levels in domestic settings are lower than those tolerated in industry, the period of exposure is far greater and little is known about the effects of prolonged exposure to low levels of dust.[30]

Exposure of urban populations is common especially in asbestos mining towns. In Canada, Lake Superior has for over forty years received large quantities of mill tailings which, in turn, have entered domestic water supplies. Far wider exposure has occurred through the widespread use of asbestos cement pipes. In the United States over one-third of all piping used for water distribution contains asbestos.[31] Cement is vulnerable to attack by acids and sulphates, and where water is corrosive the potential exists for asbestos fibre to be released. Studies designed to measure the contamination of water in American and Canadian towns have so far been inconclusive, although asbestos fibre has been identified in tap water in the cities of Ottawa, Toronto and Montreal. Asbestos has also been used widely in the filtering of beverages such as beer, sherry, port and vermouth, tonic water and some soft drinks. This has until recently been a common practice in the United States, Canada, France and South Africa. However, because of the low levels of contamination and the difficulty in identifying the population at risk, no studies have been concluded on the possible effects on public health. The design of such studies is further complicated by the presence of asbestos in all urban environments where fibre is released into the air from the erosion of brake linings. In the past, brake linings commonly contained between 25 and 65 per cent asbestos.[32] The industry has, however, claimed that any fibre released into the atmosphere is neutralized by the intense heat of the braking friction and, therefore, presents no risk to health. Once again the complexity of identifying subjects exposed only to asbestos from brake linings and not from any other source means that this question has not been resolved. Neither has any conclusive research been completed on the myriads of other ways in which urban populations can be exposed to low levels of fibre. The industry's assurance that there is no proof of a hazard is not reassuring, given the inability of epidemiologists to design sufficiently sensitive programmes to discover the effect against a general background of community illness. Asbestos in the environment is incapable of producing asbestosis, but the ability of small, casual exposure to increase the incidence of cancer is not understood.

Apart from asbestosis and cancer of the lung, asbestos is proven as the cause of a third and particularly deadly disease. That disease is mesothelioma. Surrounding the lung is a thin membrane called the pleura. A similar membrane surrounds the peritoneum or abdominal cavity. The pleura lies against the chest wall, and between the membrane and the lung is a lubricant which allows the lung to move as we breathe. Pleural mesothelioma forms in this membrane. The cancer moves across the surface of the pleura compressing the lung. The disease is inoperable as it spreads to envelope the entire lung surface. It is also a difficult disease to manage medically as the pleura contains nerve fibres which the tumour irritates thereby causing intense pain. There is no effective treatment for pleural or peritoneal mesothelioma and most patients are dead within twelve months of its diagnosis. The symptoms of pleural mesothelioma are chest pain, dyspnoea, a persistent cough, weight loss, anorexia, vomiting and lassitude. Peritoneal mesothelioma is characterized by abdominal pain, constipation and weight loss. In both forms the disease is difficult to diagnose, and in the past it may have commonly been misdiagnosed as primary lung cancer from which it is now understood to be a separate and distinct disease.

Mesothelioma is invariably associated with exposure to asbestos, and only a small number of identified cases cannot be traced to that origin. The disease is known to be more common among males than among females, which is consistent with the predominance of males in asbestos mining and manufacture. It has long been assumed that blue fibre is far more likely to cause mesothelioma than are white or brown asbestos because of its thinness and length. The blue asbestos mined at Wittenoom in Western Australia is shorter and thinner than are the fibres from the Transvaal, and this could help to explain the higher recorded incidence of mesothelioma among Wittenoom workers.[33] Within the general population, mesothelioma is an extremely rare disease and is unique in the period between exposure and diagnosis. The usual latency period is from thirty to forty years. This wide gap in time is one of the reasons it took so long to distinguish mesothelioma from cancer of the lung and to identify the connection between the disease and exposure to asbestos.

The earliest work on primary neoplasms of the pleura was published by Rabin and Klemperer in 1931, but they found no connection between the disease and asbestos. Further case studies were published in 1947 and 1953 which, in retrospect, suggest a clear association with the mineral. The role played by asbestos was firmly established in a South African study published in 1960.[34] This work by Wagner, Sleggs and Marchand was based on an analysis of thirty-three cases of mesothelioma with all but one having a probable exposure to Cape blue asbestos. The majority of the subjects had lived for a time at Asbestos Hills, which lies in the west of the Kimberley region. Asbestos has been mined in the region since 1893 and, in the early days, the fibre had been cobbled by hand with much of the work being done by women and children. With the expansion of the industry most of the labour continued to be carried out by blacks who tended to live close to the mine sites. Wagner comments in his work that it is common for the children of miners to play in the asbestos dumps.[35]

The Wagner study was researched over a period of four years, with the first case having been identified in Feburary 1956. That case was of a mine labourer who died at age 36, and it was the presence of asbestos fibres in his lungs which alerted Wagner and his colleagues to the possible role of asbestos in mesothelioma, a disease otherwise rare in South Africa. A further ten cases were referred to Wagner from a hospital which served patients in a large asbestos mining area. The major obstacle confronting researchers was the presence of non-miners within the group. Some of the patients were housewives, some were cattle herders, farmers and domestic servants who had never entered a mine. Case 4 was that of a 56-year-old social worker whose only direct exposure to asbestos was as a child. Case 15 was that of a female, aged 42 years, who had lived near a mine. Her father, a miner, had also died from mesothelioma but the woman had never entered a mine. In all, eighteen of Wagner's cases were people who had been born near or lived as infants close to an asbestos mine or mill. Eleven of these people acknowledged having been exposed to asbestos dust. The most remarkable features of Wagner's study are the hazard presented by environmental exposure and the extraordinary latency period for the disease which in most cases was well over thirty years from the date of first exposure until diagnosis.

Three years later Wagner published a second study which reinforced the conclusions drawn in the earlier work about the environmental hazard. He also found no connection between the severity of exposure and the presence of the disease. Mesothelioma appeared to differ from asbestosis in being unrelated to dose, and even the most trivial exposure seemed sufficient in a vulnerable person to provoke the disease.

It is known in the United States that more than one million people, both men and women, have at one time or another worked regularly with asbestos. Among this group there are expected to be in excess of two hundred thousand deaths from lung cancer and more than fifty thousand deaths from mesothelioma.[36] But these estimates are for occupational exposure alone, and they ignore the clear warnings given in Wagner's work published more than a quarter of a century ago that asbestos in the environment is a threat. According to an IARC study released in 1976 this threat is very real. The authors of this report go on to argue: "Indeed, no population exposed to the industrial use of asbestos has escaped this hazard, it is also likely that every country in the world has experienced this public health problem."[37] Recent medical evidence from Britain makes this threat appear all the more menacing. A case published in the *British Medical Journal* in June 1984 suggests that even the most trivial exposure to asbestos can cause mesothelioma.[38] The case is that of two sisters who lived in adjacent caravans and both of whom developed mesothelioma. Their only exposure to asbestos had taken place many years earlier when they had cleaned an outhouse roof made of corrugated white asbestos cement to remove moss. This was done with wire brushes and after a morning spent on their hands and knees the sisters were pleased that the roof had been made "lovely and white".

In judging the behaviour of asbestos producers in Australia, as elsewhere, it is necessary to establish the point at which the industry was faced with unavoidable knowledge about the dangers posed by its plant and products. It is tempting to ascribe to earlier times the present levels of knowledge about toxic substances and then to berate an industry for failing to recognize what in hindsight appear as clear warning signs. Such an approach ignores the complex way in which medical knowledge is produced and the vagaries of the process by which one medical orthodoxy is replaced by another. In every medical controversy a change of orthodoxy leaves behind a group or faction of true believers who refuse to accept that a change of belief is necessary. This has been true in the health scandals involving cigarette smoking and cancer, pesticides, vinyl chloride and any other number of carcinogens or teratogens. This clinging to traditional views has, of course, been exploited by industry to excuse itself from responsibility for the illness and death caused by its products to consumers and employees. Producers of toxic substances are all too willing to argue that medical evidence is not conclusive until the process of controversy has completely ended and a new orthodoxy has gained universal acceptance. Such an approach, whatever its morality, ignores the ways in which medical research functions. Similarly, a single article warning of the possibility that asbestos is cancer producing is not proof that asbestos should be removed immediately from the market place. Such a decision is, after all, a political matter which is not made by the medical profession itself. In the case of Australia there is a further problem: because the early breakthroughs had been made in the United States, Britain and South Africa, it is difficult to identify the exact points at which Australian physicians became aware of the asbestos hazard.

It is known that the work of Merewether and Price reached Australia soon after its publication in 1930. In 1932 the *Medical Journal of Australia* carried an abstracted article by J. V. Spark on asbestosis in which the author points out that the progress of the illness is not altered by removing the patient from the source of dust.[39] There were, over the next fifteen years, a scattering of articles in Australian medical journals which, like Spark's piece, prove that the local medical profession was abreast of developments in Britain which had distinguished asbestosis from other forms of pneumoconiosis. That growth of interest reached a high point with the Third International Conference of Experts on Pneumoconiosis which was held in Sydney in February and March 1950. The conference was sponsored by the ILO and the opening addresses were given by Sir Earle Page and Mr Harold Holt. Previous conferences held by the ILO had taken their title from the term "silicosis", and the change at Sydney heralded among other things an awareness of the individuality of the disease caused by asbestos dust. The aims of the conference, however, had not been altered, and its avowed purpose was for the exchange of information to help in the prevention of disease and in the setting of standards for exposure in industrial hygiene.

Delegates to the conference included representatives from the Commonwealth Department of Health and one can only assume that organizations such as James Hardie and CSR, both of which were involved at the time in the asbestos industry, would have taken some interest in the proceedings.

The range of papers presented at the conference was extremely wide and covered coal and gold mining, as well as discussion of the setting of standards and the best means for dust suppression. Some of those papers were of immediate interest to local asbestos producers, because they gave an insight into industrial hygiene practices in other mining sectors. An address by K. G. Outhred provides a history of pneumoconiosis in the Western Australian goldfields and, in particular, explores the work of the Commonwealth Health Laboratory at Kalgoorlie.[40] Outhred explains that pneumoconiosis had been identified as a major health hazard by

1911 when a Royal Commission found evidence of fibrosis in more than 20 per cent of miners examined. In 1925 the Commonwealth Health Laboratory at Kalgoorlie carried out clinical and radiological surveys of some three thousand subjects, and it was found that 16 per cent of the men had silicosis. Since that time the introduction of careful pre-employment testing had been successful in excluding from underground mining all vulnerable workers, especially those with a history of tuberculosis. In the period from 1928 to 1950 over 34,000 examinations were carried out on prospective miners which resulted in a rejection rate of around 10 per cent. It was also accepted practice that all underground workers should be given an annual medical examination.

At the 1950 conference, much of the discussion time was devoted to the issue of preventive measures and the best means for setting standards to protect workers from lung disease. In one such discussion, following a paper on pneumoconiosis among New South Wales coalminers,[41] Gordon C. Smith made the following comment on the significance of the coal-dust standards operating in the state: "This standard was not fixed because of any available evidence that the particular dust concentration was safe or of negligible significance in the production of pneumoconiosis, but rather it was determined from dust tests carried out in all coalfields and was considered by the committee to represent the lowest practicable limit of dustiness then attainable by good ventilation and proper use of water."[42] Clearly, state authorities in New South Wales were willing to acknowledge the economic realities which determined the setting and implementation of standards. There is no reference to abstract principles about acceptable risk and the incidence of disease. The standard was determined through a trade-off between technical efficiency, acceptable cost to the producer and the existing dust levels found in the industry. This does not mean that the delegates to the conference took their responsibility for creating a safe work environment lightly. There was considerable discussion about the question of safety equipment. The conference adopted unanimously a proposal by Dr Orenstein on the issue of respirators and other such devices. The proposal established the principle that

" . . . no reliance should be placed on an appliance which depended on the individuals using it correctly, unless it was quite impossible in the circumstances to use any other".[43] Devices such as respirators are a poor substitute for the prevention of dust at its source, and all delegates were in agreement that secondary measures could not give the safety which prevention affords.

The Third International Conference in Sydney was held shortly after the publication of Merewether's study establishing the association between asbestos and cancer of the lung. Inevitably this issue was raised in discussion and, although the lack of final clinical evidence was mentioned, there was also some concern expressed as to the seriousness of Merewether's findings.[44] It can only be assumed that local manufacturers of asbestos products, such as James Hardie, were as concerned as were the delegates about the implications of *the* Merewether report for the health of workers in the asbestos industry.[45]

Australian understanding of asbestos-related disease has, for the most part, been reliant upon research carried out in Britain, South Africa and the United States. This is true of cancer of the lung and asbestosis, both of which were first identified in Britain long before independent research had been done in Australia. But in the case of mesothelioma, Australia has the dubious distinction of being one of the chief manufacturers of medical knowledge. In 1962 an Australian physician, Dr J. C. McNulty, who is presently Commissioner for Public Health in Western Australia, published one of the first clinical profiles connecting mesothelioma with exposure to asbestos.[46] McNulty had been a regular visitor to the crocidolite mine at Wittenoom in the northwest of the state, and his article helped to validate the pioneering study by Wagner in South Africa. McNulty's subject was a white male who had worked at Wittenoom from 1948 until 1950. He was one of only three cases of mesothelioma identified in Western Australia at that time, and his case was unusual in the brevity of his exposure to asbestos. This case which arose during the latter half of the mine's life should, in conjunction with the publications by Merewether in 1930 and 1948, have been sufficient to warn asbestos producers in Australia about the dismal future faced by so many of their employees. Wagner's work had already shown that environmental pollution can cause mesothelioma. Therefore, the population at risk at Wittenoom included the families of miners, as well as those employed in the mill and the mine itself. Because of Wittenoom, Western Australia was destined over the next twenty-five years to become a fertile ground for research into mesothelioma.

In Australia during the 1960s there was considerable interest in occupational health and in diseases of the lung. This interest culminated in the First Australian Pneumoconiosis Conference, which was held in Sydney in February 1968. The conference was sponsored by the Joint Coal Board and the Commonwealth Department of Health, and it attracted delegates from all Australian producers of asbestos products. James Hardies' representatives included Dr S. F. McCullagh and Mr J. Winters, both of whom played an important role in the management of industrial hygiene at the Baryulgil mine. There were also representatives from CSR, which two years earlier had closed down its subsidiary Australian Blue Asbestos, and Wunderlich, which had been James Hardie's major competitors in the New South Wales market.

Like the 1950 Pneumoconiosis Conference the variety of topics discussed was broad, with subjects ranging from dust control and sampling techniques to papers on disease in the coalmining industry and ventilation practice at Broken Hill. A delegate from the South African Chamber of Mines, Mr P. H. Kitto, gave an impressive paper on asbestos sampling and dust control.[47] Kitto outlined the then current dust standards for the South African asbestos industry which were set at 45 fibre per cc. There was also, he explained, a South African environmental standard which set the permissible level of asbestos for residential areas at 1 fibre per cc.[48] Mr Kitto's paper and its optimistic portrayal of conditions in South Africa should, however, be treated with some scepticism, given the predominance of black African labour in the Republic's mines and the complete absence of legal rights for such workers under apartheid. Even so, it is significant that South African industry and government was willing to recognize the environmental hazards of blue asbestos fibre more than twenty years before the issue was taken up by public authorities in Australia. In retrospect, that tardiness does local authorities little credit, especially when their attitude is compared with that taken in a country whose policies on civil rights and labour relations have made it a pariah within the international community.

During the conference the issue of asbestos in Australia was raised in a paper by Dr G. Major of the University of Sydney,[49] who commented on the concern being felt internationally about disease.[50] Major pointed out that the current standard in the United States for exposure was more than fifty years old and, therefore, inadequate in light of recent research on cancer and mesothelioma. He also suggested that existing dust-measurement techniques in Australia were deficient because of variations between shifts and the variations that occurred even within specific parts of a single plant. His comments must have been of interest to the representatives from James Hardie who were present and who would have had access to the conference papers which were published in two bound volumes.

In comparing the 1950 and 1968 conferences a number of particular features are obvious. At each conference, much the same range of subjects was discussed and many of the issues raised in 1950 were still of concern to delegates eighteen years later. Changes had occurred, however, in the technology available in preventing pneumoconiosis. Those techniques, combined with a concerted effort made by industry and by government authorities, had succeeded in reducing casualties in coal and goldmining. The figures quoted by the Joint Coal Board in 1968 show a dramatic improvement following the establishment of regular, compulsory health checks of all miners and the introduction in 1943 of a dust standard by the New South Wales Health and Mines departments.[51] This achievement and the earlier eradication of pneumoconiosis at Broken Hill suggest that in Australia at least there were no problems of a technical nature in controlling dust and protecting miners from diseases of the lung. At Broken Hill, management and the trade unions were so successful that in the period from 1926 until 1968 only four cases of pneumoconiosis had been detected.[52] This makes the presence of silicosis or asbestosis in New South Wales in the period after 1945 quite unacceptable unless, of course, some evidence of a purely technical nature can be cited to explain why asbestos mining should be exempt from the standards which had been achieved elsewhere in the state. James Hardie, which operated the mine at Baryulgil in the period from 1944 to 1976, has not presented any such evidence in defending its industrial hygiene practices.

In the period since 1969 Australian medical researchers have published a number of studies on the asbestos industry and on those workers outside the industry who have been exposed to fibre. These studies include reference to asbestosis and mesothelioma and they provide a guide as to local awareness of the mineral's hazardous nature. A study by James Milne in 1969 of pleural mesothelioma among asbestos workers in Victoria suggests that environmental exposure alone is sufficient to produce the disease. One of Milne's subjects had not worked in the industry, but her father may well have introduced the disease to her through his asbestos-impregnated overalls.[53] Another study was carried out in two Sydney factories more than a decade later which reinforces the dangers of non-occupational exposure pointed out by Milne.[54] In this report by Dr I. Young, many of the subjects' only source of exposure had come from the insulation wound around steam and hot-water piping. In the second factory the source of exposure was from pollution emitted from a nearby asbestos cement plant. The results of Young's study indicate the presence of a threat to the health of the workers involved. These results were reinforced by research carried out on respiratory morbidity among Brisbane waterside workers during 1981.[55] This study, led by Dr Charles Mitchell, found that half of a sample of ninety-eight workers complained of breathlessness and fifteen subjects displayed symptoms consistent with asbestosis. Until as late as 1970 all asbestos arriving at Brisbane wharfs was packed in hessian bags which freely leaked fibre from their loose weave. Mitchell's research establishes the presence of a hazard to workers handling this material over the previous fifty years. But he is not able to even guess how many men may have contracted asbestosis, lung cancer or mesothelioma as a result.

Asbestos has been used in various Australian industries and, if the number of those workers who have also come in contact with the mineral in their lives as consumers is included, the total number of people exposed to the threat of asbestosis and cancer is increased still further. It is probable that most deaths in the past from asbestos-related disease have gone undetected because of the difficulty of diagnosis, the lack of public concern about occupational health and the latency period typical of asbestosis, mesothelioma and lung cancer. The statistics which appear in Health department publications underestimate the actual incidence of disease. These estimates tend to be based soley upon death certificates in identifying the cause of death, thereby further depressing the apparent effects of asbestos on community mortality rates. In a study published in September 1983, Robert Barnes[56] notes that the Dust Diseases Board of New South Wales (DDB) ignores any other than the primary cause of death in compiling data, even where a patient has been in receipt of an asbestos-based disability pension, and where asbestosis has clearly played some part in the death process. Despite these limitations, Barnes still managed to unearth 197 cases in New South Wales in the period from February 1969 to the end of December 1983 which had been accepted by the DDB. Of these, 29 were identified as asbestosis, 67 as bronchiogenic carcinoma and 101 as mesothelioma. Most victims had been employed for some years in the industry although, in one case, a man had been exposed in handling blue asbestos bags for less than one month. Barnes notes that there are various dangerous practices associated with asbestos manufacturing, and that during World War II crocidolite had often been wrapped around welding rods to make them burn more slowly. These rods had been used at the Vickers–Cockatoo Dockyards in Sydney, and they gave rise to very high levels of airborne fibre.[57] Because of such practices in the past, Barnes predicts that the prevalence of asbestos-related disease in New South Wales can be expected to increase over the next three decades. He is also certain that the number of deaths caused by asbestos will never be known. Even in those cases where the origin of the disease is identified and the victim is successful in receiving compensation, the trauma of chronic illness and a painful death is the grim legacy of an industry which flourished in Australia for more than half a century.

The difficulty of establishing a diagnosis of asbestosis or of proving that an individual case of mesothelioma or lung cancer is due to asbestos have not disappeared with the passing of time. It has taken many years to prove the connection between exposure to fibre and illness among workers and those exposed outside the workplace. In most industries, more than one type of fibre is employed, so that workers may be exposed simultaneously to crocidolite and chrysotile. Levels of exposure which occurred thirty or more years ago are rarely known and there has, in most cases, been a complete absence of reliable data on exposure levels.[58] Frequently, exposure levels in the past are nothing more than crude estimates, based on extrapolations from the prevalence of disease among workers in the present. This method ignores the sub-clinical effects of exposure which, although neglected, nonetheless erode the quality of life of the victims.

The problems in establishing a dose–response relationship between exposure and illness have been exacerbated by the paucity of data on dust levels. Before 1970 there was little if any data on fibre levels as distinct from dust counts which did not distinguish between dust from the host ore, in the case of mining, and the actual density of airborne fibre. Until the late 1960s in Canada and the United States, the Midget Impinger was the standard instrument used in asbestos plants. Even with the introduction of more sophisticated technology, the problem of identifying fibre levels has remained. The newest technologies such as Light Microscopy and Electron Microscopy are each fallible. In the first case, the instrument cannot detect the smallest fibres which may well be the most dangerous. Electron Microscopy is hampered by the high costs of operation and by its slowness in processing samples. This method does detect small fibres, but each sample takes several hours to process. The high magnification means that only a fraction of an individual sample can be viewed at any one time. This forces researchers to rely upon so-called representative sampling. There is no guarantee that any individual sample taken from the workplace is truly representative. Airborne asbestos fibres are rarely distributed evenly.[59]

The industry has always claimed that these new technologies provide complete protection for workers. In truth, the new technologies may have improved upon past instrumentation for measuring risk, but methods such as Electron Microscopy resemble their predecessors in giving nothing more than an approximation of hazard. The decision to declare a work environment safe is in truth a political decision based upon imperfect scientific knowledge.

Epidemiological research into asbestos-related disease has been handicapped as much by the nature of the disease as by the prevalence of asbestos in the work and community environments. With diseases such as mesothelioma, in which forty years may separate the date of exposure from the date of diagnosis, researchers face insurmountable problems in seeking to quantify exposure levels or even in establishing that exposure has occurred. The length of the latency period is also believed to eliminate many workers who, in the past, experienced high levels of exposure but who, subsequently, died either from undetected asbestos-induced disease or from some other cause. In constructing epidemiological studies of the asbestos industry, it is difficult to identify a control group of subjects proven to have had no exposure. Any person living in an urban industrial environment will have absorbed into their lungs, from any number of sources, some asbestos fibre. The consequences of low levels of exposure are little understood. In the case of mesothelioma, studies from France, the Netherlands, Italy and South Africa suggest that mesothelioma may result from exposure for as little as a single day and yet, in 15 per cent of cases, there is no evidence of contact with asbestos. Perhaps many of the 15 per cent of cases are simply unaware that exposure had occurred.

Studies of the effects of asbestos have also been hampered by researchers' reliance upon death certification as a guide in composing mortality rates. A recent British study on the causes of death found disagreement in half those cases where the clinical diagnosis and the autopsy findings were compared. The figure rose to 60 per cent when the comparison was made on those factors which contributed to death.[60] In Britain, deaths from cancer of the lung do not appear as part of the data on industrial diseases, although deaths certified as being due to asbestosis and mesothelioma do. This does not mean, however, that the official figures on asbestosis are themselves reliable. Under the existing British system it is very difficult for a worker to claim successfully for a benefit, and two out of three applications for mesothelioma fail. In the Federal Republic of Germany applicants are more successful in part because, since 1940, German occupational health boards have, even in the absence of asbestosis, accepted cancer of the lung as a compensatable disease.

Physicians treating suspected asbestosis patients are forced to rely upon diagnostic techniques which, like the instruments used in monitoring fibre levels, are falliable. Even experienced practitioners cannot be certain that their reading of X-rays is correct and that an incipient case of asbestosis has not been overlooked. Dr McNulty, who has played such an important role in public health in Western Australia, made a number of comments on this subject at the 1968 Pneumoconiosis Conference in Sydney. Dr McNulty recalled cases in which a miner had displayed the full range of symptoms associated with severe asbestosis including clubbing of the fingers and toes only twelve months after showing a normal X-ray.[61] The same margin for error is true of lung function testing, which cannot distinguish between asbestosis and other pulmonary dysfunctions. These techniques can give nothing more than a guide to the presence of disease. The only unambiguous diagnostic proof is obtained from autopsy.

The current level of medical knowledge about asbestos-related disease is fragmentary and incomplete. Little is known of the role played by asbestos fibres in cancers, other than mesothelioma, and little is known as to why asbestos provokes mesothelioma in some people, but not in others. The question as to the variation between different types of asbestos and cancer is largely unanswered. In combination, these gaps in knowledge mean that it is very difficult to prove an association or a causal relationship between a particular patient's illness and exposure to asbestos fibre. This has been used to advantage by industry in defending itself against compensation claims.

Medical knowledge has grown in a piecemeal fashion, and it is only in retrospect that it is possible to identify various turning points when decisive breakthroughs were made. The accretions to medical science have not followed a linear path, despite what both the industry and its critics would have us believe. The industry relies upon a crude form of positivism in justifying its slowness to act in protecting workers from fibre. Critics of companies such as CSR and James Hardie in Australia use much the same kind of approach in seeking to show that industry failed to respond to medical breakthroughs, such as the Merewether study of 1948 or Selikoff's work of 1964.

Both critics and the industry alike refuse to recognize the complex way in which medical research has evolved. It is that complex path which defines the levels of moral responsibility that must be borne by the industry. It is not sufficient merely to examine the publication dates of studies such as those of Wagner and Merewether in judging the behaviour of producers and public authorities. To do so implies that the industry had no independent resources or was under no moral obligation to monitor changes in the oral tradition among researchers, which most certainly moved ahead of substantive published results. The industry had access to resources and to data denied independent researchers. One can only assume that physicians employed by firms such as Johns-Manville and James Hardie would have known more than other researchers if only because of their strategic position. They knew who was working in the industry; they had access to the health records of those workers; they knew which parts of the productive process were the most dusty; they had access to dust-count data, however imperfect; and they had access to oral evidence as to the fate of male and female employees who died from respiratory disease. This was true, particularly, in asbestos towns such as Thetford in Canada and Hebden Bridge in England. Despite these advantages, producers such as Johns-Manville and Turner & Newall have adopted the same stance regarding the question of foreseeability. Each has argued that the seminal studies by Wagner and Selikoff and Merewether did not bring about an immediate change in medical orthodoxy, but were disputed within the profession for many years after their date of publication. Therefore, there was no onus upon the industry to adjust its hygiene practice in accordance to what was mere supposition. What the industry's spokesmen chose to ignore, however, is that this notion of medical orthodoxy is in effect a political concept which they have used ruthlessly to diminish their level of responsibility.

Physicians working in London hospitals in the 1950s were well acquainted with an oral tradition which portrayed asbestos as a potent carcinogen. By 1955, Merewether's report was already six years old, and the first cases of cancer of the lung from asbestos weaving and fabricating industries were becoming visible. The industry refused to act, and in its defence now argues that there was no necessity for change because government authorities were satisfied that existing precautions were adequate. A more pertinent question to ask, however, is whether such inaction can be justified in light of the industry's own level of knowledge about the dangers of its products.

The same question can rightfully be asked of the producers of various known carcinogens such as 2,4,5-T, vinyl-chloride and PCBs. The asbestos industry is not unique and its behaviour since the emergence of public awareness about the dangers of its products has been entirely predictable. It has sought to blame the imperfections of medical knowledge, the incompetence of government bodies and the physiologies of the victims themselves. It has never, however, itself sought to accept responsibility. The industry has been shielded in this calling to account by the conservative character of the medical profession and by that profession's avowed apolitical role in the settling of public disputes. By their own admission, medical practitioners in this country, as elsewhere, share a number of assumptions as to how their profession should function. They also share various characteristics in terms of social origins and ethnic background, and material and ideological interests which they are less comfortable in discussing. Neither medicine as a science nor its research products is immune from the ideological presuppositions of its members, or of those who gain material advantage from the profession. Medical knowledge about asbestos has throughout this century been political knowledge and, if physicians have chosen to ignore this truth, their ignorance has not lessened the influence of political forces in the production and dissemination of research. This process is seen most clearly in the setting of standards for exposure in Britain, the United States and Australia.

The major international conferences held in New York in 1964 and at Lyons in 1972 were convened in response to growing public concern about the carcinogenic effects of asbestos. Both conferences were political events which had an immediate impact upon the industry. Both conferences were influential in changing the attitude of government authorities to acceptable dust levels, although the rate of change proved to be slow and erratic. The asbestos scandal of the past fifteen years is not the first but the second time that this drama has been played out in the public arena. During 1929 and 1930 the American asbestos industry suffered a rash of litigation when more than a hundred thousand individual claims were lodged for asbestosis and silicosis disease.[62] Although the asbestos industry was not the major target of these claims, the industry was well warned of the consequences of maintaining and operating unsanitary mines and plants. In the United States until the 1930s, the amount available for occupational disease compensation was very small, and "eating dust" was an accepted part of life in any mining town. It was usual for individual firms to pension-off diseased workers, and it was not until 1929 that the first claims for asbestosis were filed.[63] A high percentage of women were employed in the industry which helps to explain why the first major health studies of asbestos workers were not completed until as late as 1935. In America, as in Britain, most of the worst exposure occurred in the textiles industry, where it was a serious disease. This view was held by Lanza who had a close relationship with the industry for whom he acted as a consultant. But Lanza was confident that the disease could be eradicated for, as he commented, "As soon as the hazard was realized, industrial firms fabricating asbestos took energetic steps to control the dust so that it is probable that cases of asbestosis will become uncommon."[64] Like his contemporaries, Lanza was unaware of the threat posed by cancer and mesothelioma, from which a reduction in the incidence of asbestosis would never guarantee immunity. In any case, Lanza's predictions about the incidence of asbestosis proved false largely because the industry lacked a commitment to improve work conditions.

Lanza's classic study on asbestosis and silicosis was published in 1938. In that same year the American Conference of Governmental Industrial Hygienists recommended a level of five million parts per cubic foot as the maximum permissible level of exposure to asbestos. This soon became an unofficial standard throughout the industry. The standard was intended to prevent asbestosis and the James Hardie organization claims to have adhered voluntarily to it at Hardie plants throughout Australia. Until the introduction of an official standard, this practice continued for more than twenty years. In Britain, the 1931 asbestos regulations remained unchanged for over thirty years, and it was not until 1969 that a more stringent standard was introduced. In the United States, the question of an asbestos standard has, especially over the past fifteen years, gone through various seasons, as public pressure has forced government agencies to react against the industry. In 1975, the United States Occupational Safety and Health Administration (OSHA) proposed that the existing standard of 2 fibres per cc be lowered to 0.5 fibres per cc. In December of the following year, the National Institute of Occupational Health and Safety suggested that the standard be lowered even further to 0.1 f/cc, which was at that time the lowest level detectable. The standard accepted by the OSHA in 1981 was 2 f/cc to be measured over an eight-hour period. It was also recommended that no worker be exposed at any time to more than 10 f/cc. This standard existed until 4 November 1983, when the OSHA issued an emergency temporary standard to reduce the permissible level from 2 f/cc to 0.5 f/cc. Although this emergency order was challenged and subsequently revoked by the Federal Appeals Court, the damage had already been done to the industry and to the credibility of its commodities.

In July 1984, the OSHA began a series of public hearings into safety standards which again took up the issue of lowering the level from 2 f/cc to 0.2 f/cc or even to 0.05 f/cc. Unfortunately, the OSHA proposal emphasized the use of respiratory equipment as distinct from dust-prevention measures which the industry has fought against so trenchantly.

The OSHA standard was based on an abstract count of the number of excess cases of cancer which can be expected to arise from varying levels of exposure. In the case of an exposure to 10 f/cc, weighted over a 45-year period, the OSHA expects that there would be 165 excess deaths per 1,000 subjects from cancer. At a level of 1 f/cc the number would fall to 64 excess deaths per 1,000 subjects. The advantage of the OSHA method is that it recognizes the interaction between the economics of the industry in accepting a particular standard and the consequences that arise in terms of the damage done to individual lives. The abstraction involves so many dead employees, who also happen to be fathers and husbands and sons and brothers. It has never felt comfortable with the kind of approach favoured by the OSHA. It prefers to treat safety standards as involving a trade-off between economic feasibility and existing technical choice, both of which prevent capital from creating absolutely safe work environments.

The history of asbestos mining and manufacture shows that the setting of standards for acceptable levels of exposure is the outcome of a variety of economic, political and technological factors. Standards are never merely technical devices guaranteeing workers immunity from occupational disease, even though governments may seek to portray them in this light. Above all else, standards are a reflection of political realities.

In 1968 the first steps were taken in Britain to change the existing standard which had survived for more than thirty years. In that year, the British Occupational Hygiene Society claimed to have discovered a dose–response relationship for asbestos fibre which allowed the risk of exposure to be quantified accurately for the first time. Following the 1964 New York conference, the British Occupational Hygiene Society set up a subcommittee to evaluate the existing legislation. That committee included representatives from the industry, from universities and from government departments. The weight of technical expertise on the committee, however, was dominated by the industry's representatives who had a level of experience not shared by the universities or by the government departments concerned.[65] The committee based its final recommendations on a single piece of research carried out at a Turner Brothers asbestos plant at Rochdale, near Manchester. The committee recommended a level of 2 f/cc which was presumed to give a 1 per cent risk to a man working in the industry for a fifty-year period. This risk level was based upon a calculation of the chances of contracting asbestosis. It ignored the dangers already well established in the medical literature of mesothelioma and bronchiogenic carcinoma.[66] From the industry's point of view, the 2 f/cc standard was eminently acceptable because it was the standard already in operation at Rochdale and at many other plants throughout the country.

The committee's recommendation was accepted by the British government and in the following year it became the official standard. The only benefit to emerge for employees and consumers was the increased level of public debate which the inquiry stimulated about the dangers from exposure to asbestos.

In March 1974 the Chief Inspector of Factories for Britain announced that during the past year there had been 128 recorded deaths from asbestos-related disease. The industry welcomed the release of such figures as demonstrating its success in ensuring the health and safety of employees. Mr J. H. Dent, Chairman of Cape Asbestos, explained that for every dollar spent on new plant the industry also spent a dollar on dust-extraction equipment.[67] Similarly, for every horsepower of energy expended in driving plant the same cost was expended on freeing air of fibre. At the time, the British industry was dominated by two companies: Turner & Newall and Cape Asbestos. Both corporations held extensive mining leases in Canada and South Africa and both claimed to follow the same hygiene practices in their South African holdings that were mandatory in liberal democracies. In the Republic, of course, civil liberties and the occupational health of black African workers are of little importance.

In Britain, the standard has been successfully improved. The major change in British policy came, however, in 1979 with the publication of the Health and Safety Commission's final report.[68] The report acknowledged that the 1969 review had been effective, especially in the phasing out of blue asbestos imports. The commission now proposed a total ban on blue asbestos, as well as on all products containing that particular fibre. It also proposed that, as from September 1980, the standards for chrysotile be further reduced to 1 f/ml. Perhaps the most encouraging aspect of the report was the recommendation that the British government press for similar tighter controls in the EEC, and by the WHO and the ILO.[69] British industry protested against the changes which it estimated would cost asbestos producers £18.7 million initially, and a further £10 million each year thereafter.[70] The producers were not willing to concede that the changes were long overdue and would save many lives. Turner & Newall and Cape Industries had, in fact, fought a long and expensive campaign over the preceding five years to block the recommendations which the commission eventually introduced. The industry was well aware of the long-term consequences for its products once the standard had been lowered again.

In 1983 the Health and Safety Commission imposed a stricter control on the uses of all types of asbestos. It also recommended a ban on the import of blue and brown fibres which was instituted in August 1984. In the case of chrysotile the accepted level was reduced from 1 to 0.5 f/cc, and blue asbestos exposure remained unchanged at 0.2 f/cc.[71]

The strategy used by asbestos manufacturers against their critics in North America and in Western Europe (the EEC states) has been largely the same. The industry has fought at every turn the introduction of more stringent standards in the workplace. In so doing, producers such as Turner & Newall and Johns-Manville have introduced the concept of the "susceptible worker" which is identified as the cause of all disease. This concept is based upon the fact that not all those who are exposed to asbestos become ill, just as not all people who smoke cigarettes die of lung cancer. All biological organisms, including humans, vary in their response to carcinogenic agents. Through a variety of factors including genetic makeup some people may be more vulnerable than others to a particular carcinogen. The industry claims that it is not feasible to invent work environments that are safe for these vulnerable individuals. It is far better, according to spokesmen from Johns-Manville, to assign such people to less hazardous employment than to try and construct a workplace

that is safe for all. An analogous strategy has seen industry concentrate upon the role of cigarette smoking as the cause of lung
cancer among asbestos workers. It is smoking, rather than exposure to fibre, which is therefore seen as the major problem. The
answer is not to reduce fibre levels in the factory, but to discourage smokers from entering the industry, or to induce asbestos
workers to desist from smoking.

These kinds of strategies have the advantage to producers of
shifting the blame for illness from the work environment on to
the shoulders of the victims who are seen, to some extent, as
culpable. Once the concept of the vulnerable worker is accepted
then the worker is called upon to be segregated. The immediate
effect of this line of reasoning is that the workplace is left intact
and the employer is exonerated from all responsibility for injury
caused to employees. The most obvious flaw in this argument is
found in the environmental fallout from carcinogens, where
vulnerable communities cannot be segregated from carcinogenic
commodities. Manufacturers never argue that consumers, like
employees, should be subject to tests as to their fitness to come
into contact with certain products. Corporations such as James
Hardie, Johns-Manville and Turner & Newall may be content to
promote the exclusion of vulnerable people from the workplace,
but they prefer to ignore the equally reasonable solution of
excluding their goods from the market place where vulnerable
people also live and consume their products. There is good reason
to believe that consumers are in need of protection from corporations that are willing to employ such arguments in promotion
of a bizarre vision of the public good.

4

Wittenoom: The Mine

Wittenoom lies in the heart of the Pilbara region some sixteen hundred kilometres north of Perth. The town itself is sited in a spectacular gorge that gives Wittenoom the kind of appeal which attracts tourists to the north of Australia. What distinguishes Wittenoom from other towns is the presence of the largest deposit of crocidolite or blue asbestos to be found outside of South Africa.

Asbestos has been mined in the Pilbara since 1908 but, until 1944, mining was done only on a small scale with individual miners gouging ore and processing fibre by hand. In 1930, Lang Hancock discovered blue asbestos at Wittenoom Gorge, but it was not until eight years later, with the rise in the price of asbestos fibre, that he first began to mine the deposit. From 1938 until 1944 Hancock and a partner, Albert Wright, worked the site mainly tributing asbestos from individual miners who were paid according to the amount of fibre they recovered. Processing was done with a primitive crusher, a hammer mill and a screen, and the quantities returned were small. Hancock made as much money from shipping supplies and alcohol in the back-freight from the coast to the mine as he did from asbestos. This situation changed, however, during the later years of World War II, as the price of asbestos began to improve. At that time the sugar conglomerate, CSR, had become interested in entering the building materials industry. The Colonial Sugar Refineries already had some experience in operating a small white asbestos mine at Zeehan in Tasmania, and it approached Hancock and Wright about the development of the mine at Wittenoom. As a result of those negotiations, Australian Blue Asbestos (ABA) was formed in 1943 when CSR purchased 51 per cent of the lease. For the next five years Hancock and Wright managed the operation, and CSR provided the finance and most of the technical expertise. In 1949 the company became a fully owned subsidiary of CSR when Hancock and Wright sold their 49 per cent share in the venture. From that time until 1966 when the mine closed, Wittenoom was the largest asbestos mine in Australia, and it produced blue fibre for both the domestic and export markets.

Lang Hancock's involvement at Wittenoom did not end in 1949, for it was Hancock who purchased the mine, the treatment mill, the Wittenoom Hotel and various other administrative buildings from CSR in 1966. The purchase, which cost more than $1 million, was motivated primarily by Hancock's plans for the development of the iron ore industry in the Pilbara and not out of any desire on his part to re-enter the asbestos industry. Because of its location and the presence of an infrastructure, Wittenoom appealed to Hancock as a possible administrative centre for his Pilbara empire. Today Hancock retains the freehold of many buildings in the town where his sister still lives. Wittenoom is no longer a bustling mining town which, during a thirty-year period, provided employment for over 6,000 men and women. This figure excludes the wives and children of miners who lived in the town and the people who worked in non-mining jobs, such as police, clerks and hotel workers. It also excludes the large number of people who visited Wittenoom as tourists.

During 1944 the total value of blue asbestos mined at Wittenoom was a mere £10,090, which gives some indication of the smallness of the mine prior to CSR's active involvement.[1] In that year a boarding house was erected and a site cleared for the homes of future miners. This work was frustrated by a shortage of building materials, caused by wartime conditions and by the isolation of the town.[2] A road from the town to Wittenoom Gorge, the site of the mine, was constructed by the Main Roads Board and, according to parliamentary records, by the end of the year a firm basis for an asbestos industry in Western Australia had already been established.[3] Apart from persistent problems in securing building materials, the mine was faced with a shortage of labour which was to last until the end of the war. Throughout the state, labour was in great demand and many mine owners had given up hope of securing skilled workers. In 1944, the lowest wage paid to an ABA employee was £10 per week, and the company was committed to erecting good-standard housing for all its employees,[4] which indicates the importance placed on attracting miners to the town. Despite the laying of sealed roads by the Main Roads Board, the community suffered from a shortage of stores and a poor shipping service, which was the blight of all communities in the northwest. Even perishable goods had to be delivered by air,[5] which was expensive and did little to attract labour.

Australian Blue Asbestos began production in July 1945 with good yields. Initial tests indicated that the fibre was of sufficient quality for use in asbestos cement sheeting, and that full-scale mining would be successful. Although the state government did everything it could to find reasons that Wittenoom should succeed, Mines department officers were not always optimistic about the future of the project. The mine was isolated, and involved the production of a material which was considered important especially in wartime. It also involved a major Australian company and, thereby, held the promise of employment in a state starved of industry. According to department files, the mine's indifferent performance during the first two years was due entirely to the effects of the war and the resulting shortage of monetary outlay and materials.[6] The department believed that these conditions would soon improve and that the success of the venture was guaranteed. And yet there are some department files which give a very different picture. According to the State Mining Engineer, the management at Wittenoom was inexperienced, and he saw little likelihood of any improvement in the mine's performance. He suggested that "the position at the moment [1946] is ominous for the company and also for the state as a whole."[7] He was, of course, referring to the asbestos industry, rather than to the mining industry or to the economy of Western Australia. The milling plant also gave cause for concern. Department records show that insufficient attention had been given to dust suppression and that the problem had been ignored by the owners. As one officer commented, "When there is no flow of air in the Gorge, which does not occur very often, dust conditions in the vicinity of the plant have been terrific." The report goes on to suggest, "In any dry crushing plant the dust menace must be accepted as a major principle of design and not as an afterthought."[8]

These rather gloomy predictions from the department are tempered by a vein of blind optimism, which is present in the most negative assessments. Even when an officer is commenting on the "terrific dust" problem, he then goes on to say that the problem will soon be overcome as the company is spending much money on improvements.[9] The Mines department, like the state government that it served, was anxious that the mine be successful. The State Mining Engineer admitted in a letter to the Secretary of Mines in March 1945 that his department was quite willing "to introduce a little propaganda" into press releases predicting the Wittenoom operation as a major success.[10] The propaganda to which he referred appeared in an article in a Perth newspaper. The article, entitled "Asbestos: World Production", claimed that the development at Wittenoom would eventually enable Western Australia to supply the whole of Australia's needs for asbestos, as well as allowing for an export surplus.[11] The author then went on to describe the high quality of the Wittenoom product and the vast reserves, which he said were sufficient to last until the end of the century. The Mines department must have been pleased with the author's comment that "the value of a sound progressive asbestos industry in the post-war settlement of the North West would be invaluable".[12] Obviously, the Mines department was from the beginning committed to the success of ABA, despite the numerous obstacles which were evident from the first days of the mine's operation.

Colonial Sugar Refineries' interest in Wittenoom began in 1941. In that year, CSR purchased an existing asbestos cement plant at Alexandria in Sydney, a plant which had relied upon imported fibre from Canadian and South African producers. From 1944 onwards, the company also operated a mine at Zeehan. It was expected that this mine would produce at least some of the chrysotile for the Sydney operation. Throughout the war years there were considerable problems in securing materials and finance for any industrial enterprise, but in the case of asbestos CSR found that government backing was granted immediately.[13] Asbestos was regarded as a strategic material and Canberra was anxious that Australia should become self-sufficient as soon as possible as a safeguard against the blockage of supplies from Canada and South Africa. One Mines department file contains the following comment on the importance of securing a local source of fibre: "At present with the war on our hands, the demand for fibre is imperative, and its source immaterial, and, as imports are limited, as much as possible must be produced here."[14]

Colonial Sugar Refineries shared a close relationship with the American conglomerate Johns-Manville and, in 1945, senior personnel from the company visited Manville mines and mills in the United States.[15] During these visits, CSR technical officers collected information about the mining and milling of asbestos cement products, which were expected to attract an expanding market once the war had ended. The company, however, had little experience in mining or in the building materials industry. And no Australian company had experience in running a large-scale asbestos mine. The existing mines at Baryulgil and at Zeehan in Tasmania were small oeprations producing only a few hundred tons of fibre each year. This lack of experience did not discourage the company from seeking help from state and federal governments alike.

The Mines department in Perth believed Wittenoom possessed one of the largest blue asbestos deposits in the world, and that its exploitation could play a vital role in the development not only of the northwest but of the state as a whole. Presumably, asbestos could provide much the same stimulus which gold had given the state fifty years earlier. For historical reasons, the Mines department formed a large and influential part of the state bureaucracy, and it was very much imbued with a passion for development.[16] Wittenoom and ABA, therefore, held great appeal at the state level. In Canberra, the site of the mine seemed equally attractive, because it promised to help in populating an isolated region seen as vulnerable to Australia's neighbours to the immediate north. In the shadow of the bombing of Pearl Harbour and Darwin, the ideology of "populate or perish" held sway over public and private imaginations alike.

During 1945 and 1946 there were close negotiations between CSR and state government departments, principally the Department of Mines. The government agreed to provide "appropriate amenities" for the town, including some public housing under the terms of a Commonwealth–state agreement.[17] The Commonwealth had already by that time undertaken aerial surveys for access routes to and from Wittenoom. The state government was also willing to provide some assistance with concessional freight rates from Port Sampson to Fremantle. These were initially fixed at a rate of ten shillings per ton. It refused, however, a company request to subsidize air travel for workers to and from the town. The company continued to apply to both state and federal governments for assistance and, in September 1947, it succeeded in attracting public support for the construction of a fuel depot at Port Sampson, and the existing freight subsidy on asbestos shipped to Fremantle was extended to include all freight sent to the mine. These arrangements followed a request from ABA to the then prime minister, Chifley, and to the state premier, McLarty, both of whom were involved directly in the decision.[18] This kind of commitment at the state and federal levels would make it difficult for any officer from the Mines or Health departments to suggest that the mines should be closed because it was unhygienic.

The Mine and the Mill

The first CSR mill at Wittenoom was constructed from the cannibalized parts of other plants. The mill building came from the Youanmi Gold Mine and the powerhouse was shipped in from Whim Creek. The crusher used to reduce the ore was originally owned by Jacques, a Melbourne firm. According to one senior CSR employee, at no time did the company seek to achieve a particular dust level; it merely wanted to keep the dust down as much as possible.[19] In this, it received no technical assistance or advice from the Mines department, despite the fact that from the very first trial run with the mill it was obvious that dust was going to be a problem.[20] The company always assumed that the plant would eventually be improved to a point where dust control was adequate, but it was not until 1956, when the mill was moved to the Colonial Gorge, that management was satisfied with hygiene. The Wittenoom Gorge mill operated for thirteen years from 1944 until 1957. From 1956 there was a brief period when the two mills, the Old Mill and the new plant, were run simultaneously. According to Cecil Broadhurst, a manager within CSR's fibre division, the major problems with the Wittenoom Gorge mill were due to the constant enlarging of the plant which was done on an *ad hoc* basis to fill higher production quotas.[21] Each addition led to further increases in dust levels, although Broadhurst could not specify what these were, because no dust counts were kept until as late as 1952 or 1953.

All mine- and mill-workers at Wittenoom were exposed to dust containing asbestos fibre. The mine was highly mechanized with the use of heavy machinery, except at the rockface where conditions were too cramped. The broken ore was fed by conveyor belt to the adjacent mill which was constructed on three separate levels.[22] The primary level contained the picking belt, where a team of six men handsorted the ore, thereby upgrading the material. The ore then moved to a crusher where some fibre was separated, with the main process of dividing the fibre taking place at the third and lowest level. In any such plant large quantities of dust are generated. At Wittenoom, water was used only during drilling at the ore face and in the scraping process: the rest of the plant was "dry". Most of the dust in the mill was the result of poor ventilation. This problem had been brought to the owner's attention in 1945, but CSR was apparently either unwilling or unable to correct the situation.[23] It is believed that much of the dust arose from the use of the major access tunnel as the primary airway.[24] All traffic, including the conveyer belt, passed down this tunnel and, although water sprays were placed at various points near the belt, they made little difference to pollution. Dust within the mill also tended to enter the mine through the main air-duct.

Once extracted, the fibre was packed into 100-pound jute bags and dispatched in grades according to length. The packing of the longest grade was done manually by workers shovelling the fibre into the bags with packing sticks. The inferior fibre was packed mechanically. All the men working in the mine and mill were exposed to some dust and airborne fibre, but the men employed directly below the conveyor belt would have to brush away the asbestos which fell on them constantly from above. The dust was so thick outside the mill that the flying doctor, Dr Eric Saint, who visited Wittenoom regularly during the 1950s, found the pollution a useful aid to navigation. A friend of Saint's recalled: "He used to fly in and see the blue dust from the mine and mill. He used to joke about it and say that the local airline used to glue this on the landscape as a homing sign for the township."[26] The story may be apocryphal, but it is consistent with oral evidence from the men who worked for ABA.

Allan Osborne was a trained metallurgist who first arrived at Wittenoom in 1950. He remembers the conditions as being poor and that the mill had obviously been put together from bits and pieces.[27] At that time the company was anxious to increase output from eighty to one hundred bags per day, that is, to a level of 8 tons. The major problem Osborne faced was in trying to reduce the high levels of stone and gravel in the bagged fibre to which manufacturers had objected. The host rock at Wittenoom was extremely hard, so hard that it would, in Osborne's words, "scratch glass". The host rock would soon wear out the ducting in the mill and after as little as two hours holes would appear allowing gravel and stone to pass through into the bagged fibre. These holes were bound with rubber patching or hessian during the week to allow work to continue, with serious repairs being carried out at weekends. In 1950, building materials of any kind were scarce and all supplies to Wittenoom had to be transported by ship and then by road across from Roeburne. Spare parts for the plant came from Sydney, which meant that major repairs or innovations would take weeks, if not months, to complete. More often than not parts were unavailable. The company, as Osborne recalled, discussed the ducting system with the Mines department and without success tried various remedies, including the use of rubber instead of metal in the piping. As far as Osborne can remember, head office in Sydney was always ready to grant money for dust collection, and he was sent to South Africa on a fact-finding tour in the hope of introducing new methods to the mine. During the years of his employment with CSR, Osborne made eight such trips but, in general, the methods used at Wittenoom were superior to those he found in plants in South Africa and Rhodesia.

The ore at Wittenoom was graded at around 5 per cent, so that 20 tons of ore would yield 1 ton of fibre. This yield improved somewhat after 1959 when the mill was moved into the Colonial Gorge. However, the most serious engineering problem at the mine was not the rate of return of asbestos from the host rock, but the very narrow seams in which the fibre appeared. The seams were at times as narrow as two-and-a-half inches.

During the first years of operation CSR tried various methods to control the dust, and experimentation was still being carried out when the move to the new mill was completed. At one time ABA sought to introduce a "wet method" in the extraction process, but this was found to damage the fibre by reducing its length, which was one of the main attractions of Wittenoom crocidolite to producers. Much of the mine's output was exported to Johns-Manville in the United States, and senior personnel from Johns-Manville visited Wittenoom on various occasions.[28] Presumably, there was some discussion between Johns-Manville and ABA about dust suppression and the best methods for improving the product. However, at no point in the life of the mine was the dust problem ever overcome, and every person in the mine or the mill was subjected to high levels of exposure. Allan Osborne, who was a member of the senior staff of CSR for more than twenty years, was transferred to Wittenoom at the beginning of 1950, where he worked in much the same conditions as the rest of the mill-hands. In 1957 his chest X-ray showed signs of silicosis and he was sent to Perth for further tests.[29] Subsequently, Osborne received notification from the Health department that he had contracted silicosis and asbestosis. Having been told this, Osborne continued on in the same job as before.[30] As always, he would visit the plant each morning, and he would go underground to inspect the mine face. Presumably, ABA was aware of the Health department's diagnosis and yet there was no suggestion that Osborne be pensioned off, or that he be sheltered from the more dusty parts of the operation. Osborne's experience is all the more significant because of his seniority with the company.

George King is another senior executive from CSR who remembers the problems of dust control at Wittenoom. King was at one time the director of ABA, and he rose eventually to the rank of general manager with CSR after having joined the firm in 1933. King first visited Wittenoom in 1948 to work on a number of technical problems in the recovery of fibre and the suppression of dust. He found that the levels of dust varied according to the intensity of the treatment process. The basic method of dust control involved the use of cyclones or centrifugal dust collectors. Air suction would be applied at the point at which the dust was being produced. The air would then be conveyed by a duct to a cyclone where the dust was collected. The company gathered information from overseas on the best methods for control, especially during the period from 1948 to 1950.[31] Even so, King recalls that in the immediate post-war years the emphasis at the mine was on higher output rather than on matters of industrial hygiene.[32] According to King, the company was, if anything, concerned about the threat of silicosis among the workforce, and it was to prevent this particular disease that efforts were made to reduce dust levels. He also remembers there was always a bluish-grey haze coming from exhaust chimneys on the mill roof which blew dust directly into the air.[33]

According to King, until the middle of the 1950s ABA was not concerned at all with the threat of asbestosis. Even though Merewether and Price's seminal report had been published more than twenty years earlier, that was a disease which in the view of ABA had not been distinguished from silicosis.[34] We do know, however, that CSR had no specialist medical personnel dealing with industrial health matters, and that such questions were left to the discretion of local managers. It was up to management to acquire whatever knowledge was necessary to ensure the safety of employees. Until 1953, CSR did not employ its own company physician.[35] The consequences of that decision are still being felt by the men who worked at Wittenoom those many years ago.

When King first visited the mine it was obvious that the plant was not going to live up to expectations. The yield was not encouraging and the host rock was extremely hard. Although the company had little experience in the mining industry, it did possess an excellent technical library which contained literature covering all aspects of asbestos production. This material was circulated throughout the organization so that local managers, in theory at least, always had access to the latest literature relating to their particular plant. The library accession list was sent routinely to every CSR factory, including Wittenoom. King remembers the accession list and the library, but he cannot recall any references to the health dangers of asbestos fibre.[36] King's testimony is borne out by a company pamphlet, presumably issued to all ABA employees and dated 1959. The pamphlet is titled *Basic Safety Rules,* and it lists rules for working in the mine and mill. It refers to the need to shield the eyes and face from dust, but there is no mention whatsoever of the need to avoid inhalation of dust and fibre.[37] Neither is there any reference to the appropriate use of respirators, or the threat posed by asbestosis, silicosis or cancer of the lung, all of which were occupational diseases at Wittenoom. According to former miners, this pamphlet was issued without change up until the closure of the mine in 1966. The pamphlet was in English and Italian, and it was unintelligible to those miners who arrived from Europe without a command of either language.

Although no official figures are available, it is believed that, excluding Britain, almost half the miners at Wittenoom came directly from European countries. The total number of people employed by ABA was 6,200,[38] of whom nearly half stayed less than three months. There were various reasons for the high turnover of labour, including the severity of the work conditions and the quality of life in the town. To recent immigrants, Wittenoom must have come as a shock with its alien landscape and its appalling isolation. Frederick Harmpell remembers being recruited by ABA when living in a village in the north of Italy. Advertisements arrived at the local employment office offering jobs in a hydroelectric scheme in Australia.[39] In his village alone, twenty-seven young men signed on for what seemed an opportunity to make good money and to travel to the other side of the world. The men were flown directly to Darwin and it was during the flight they were told of a change of destination. Some of the men were to be offered work in Tasmania and the others would have jobs at Wittenoom. Harmpell recalls that on arriving at Wittenoom many of the men wanted to return home immediately and a group actually telegrammed the Italian Consulate in Perth demanding repatriation. Most, however, stayed under a scheme whereby their wives and children were to join them after twelve months' service. Initially, Harmpell stayed for only a brief period and then worked in timber mills in New South Wales. But the money was never as good as at Wittenoom so he returned to the mine. The conditions were harsh and the language barrier made life difficult. It is not clear from Harmpell's evidence, nor from that of other miners, whether ABA did its own recruiting in Italy and Yugoslavia. It is certain, however, that some workers underwent a thorough medical examination by a company-appointed medical officer prior to embarkation for Australia.

Santo Janjetic emigrated to Australia from Italy in 1960.[40] His original destination was the Bonegilla migrant camp in Victoria, but this was changed after his ship, the *Roma,* stopped at Fremantle. An ABA official came on board and informed the men that high-paying jobs were available at Wittenoom. The company would pay the men's fare to the mine and they would be given assistance with food and accommodation. The wages, so they were told, were the best available anywhere in Australia and the conditions were also good. As Janjetic recalls, twelve men left the *Roma* and the ABA official organized their disembarkation with Immigration. When Janjetic arrived at the mine he was allowed to buy food, a mattress, linen, working clothes and shoes at the company store. Payment was to be deducted from his wages as was the airfare from Perth. After six months' service the deduction for his fare was to be refunded. It was made clear to the men that they could not leave the town until all their debts had been cleared. The English-language classes provided by the company lasted only two months, which hardly allo-

wed sufficient time for the men to become proficient. For many
of the miners, the language barrier was a handicap, and it could
hardly have helped in the practice of industrial hygiene and safety in the mine or the mill.

Igino Casale spent four years at Wittenoom from 1962 until
the closure of the mine in 1966.[41] Most of this time he worked in
the mill after an initial period underground. Casale is now 50
years old and he is employed doing light casual work. Even so,
he is constantly tired and breathless, which is unusual in a man
who has led an energetic life. Casale is one of three brothers who
worked at Wittenoom: the other two have since returned home
to Italy. He went to the mine because it was the only work he
could get at the time. The company made no mention of illness
or danger in working with asbestos, and no resirators were available. Such equipment would in any case have been useless as
there was so little air in the mine shaft that a respirator would
have made work impossible. Like the other miners, Casale never
complained to management about the conditions, neither did he
complain to the trade union, the Australian Workers Union
(AWU), which played a passive role in the town. Like many
miners, Casale worked at Wittenoom for various periods, leaving only to return again lured by the attraction of high wages.
He liked Wittenoom because of the extra money he could earn,
and as a result he managed to buy a house, a car and furniture in
Perth, which otherwise would have been impossible. Casale is
now disturbed by the number of his former workmates who have
died in middle age. There are other former miners who have less
pleasant memories of the mine and life in the town.

Zeff Panizza followed his brothers to Wittenoom, a place he
recalls where nobody cared about one another and where the
migrants were "treated like animals".[42] During the voyage from
Yugoslavia, Panizza and the other immigrants were told about
the mine and that wages and conditions were excellent. Many of
the men were under two-year contracts with the Department of
Immigration and the prospect of working in the mine for such
high wages was appealing. Before embarking Panizza had been
shovelling snow at a dam site in Yugoslavia, but when he arrived
in Perth he was put to work in a brick factory operating a kiln in

the heat of a Perth summer. The wages at Wittenoom were far higher and, besides, the work at the kiln was unpleasant, so after a brief period he signed on with ABA.

When he arrived at Wittenoom, Panizza had the advantage of speaking English, unlike many of the migrants who could not master the language. After spending the first three months in the mill, Panizza went underground where there was less dust and the work was easier. There was no light in the mine and the men were constantly plagued by huge mosquitoes. The mine was also cramped as the ceilings were only three feet high, so the miners spent most of their time bent double working on their hands and knees. After completing a shift the men would come out of the mine covered in grease and fibre which would stick to the skin. The company did provide showers but often there was not enough hot water. In the town there was the opposite problem. Because the water pipes were laid on the surface, all the water was hot, which meant that miners returning home from work would have to fill the bath and wait for the water to cool down before washing. Prior to going to Wittenoom, Zeff Panizza earned £7 or £8 a week in Perth. At the mine he earned, on average, £7 a day. The best money was to be made by working double shifts, as in those days, there was no district allowance for working the in northwest. It was common for men to become drunk and miss the night shift which gave keen workers the chance to make extra money. Panizza did particularly well out of his years with ABA because of the frugal way in which he lived. He neither drank nor smoked and he always took extra shift work whenever it was available. He did not eat in the company canteen, but cooked his own food and, on returning to Perth, he had saved enough to buy a taxi. The mine proved to be the turning point in his work life. For some of the miners, however, Wittenoom has proved less beneficial.

The cost of living in such an isolated town was extremely high, and for many of the miners and their families the high wages were not sufficient to allow them to save. In 1959 a four-gallon drum of kerosene cost twenty-five shillings, and the miners had no choice but to pay the price demanded by the company store.[43] A bottle of beer which cost three shillings in Perth sold for four times that amount at the Wittenoom Hotel. For a time, some of the men formed a syndicate, shipping beer and other supplies under contract from the coast, but the company soon put an end to the practice which threatened its monopoly over suplies coming into the town.[44] Meat was also expensive as were vegetables, although some food was grown locally. One of the miners remembers that for a time the local police sold meat, but they charged high prices which were little different from those at the local store.[45]

The problem of indebtedness began for most miners and their families the moment they arrived in the town. The company houses were completely empty and the miners would have to buy bedding, furniture and refrigerators from the store, which ran a profitable trade buying and selling second-hand furniture. The store would buy the belongings of departing miners who could not afford to freight household goods to Perth. The store paid a low price for such goods which it then resold to incoming families at a profit. In the late 1950s a second-hand refrigerator would be sold for £90 and the store held a stock of eighty such units, which it sold and resold again and again always at the same price.[46] Some of the men tried to make their own furniture but there was a lack of timber which, like everything else, had to be shipped in from Perth.

The town had a weekly newsletter called *The Wittenoom Gossip,* which carried articles and notices about weekly events. In 1957 a man named Thomas wrote an article for the paper about a giant cauliflower which one of the miners had grown in his backyard.[47] The vegetable was so huge that Thomas estimated, if sold in one piece, it would cost well over £1. Soon after the article appeared Thomas was approached by the manager of the local store, the man in fact who was selling the cauliflower. He was also approached by a mine official. Both accused him of trying to cause trouble. In writing about the high cost of living he was encouraging, so they said, claims for higher wages and he was warned that in future he should write no further articles.

The senior personnel at Wittenoom did not live in the town, but at a site some eleven kilometres away, towards the mill. This settlement was itself divided into two parts. The manager and the accountant lived in the upper houses, and the senior operating staff and the other personnel lived in the lower section. These two sections were divided by a creek. The division was indicative of the rigid social barriers present in all parts of the Wittenoom community, which was divided along ethnic and class boundaries. All amenities at Wittenoom were controlled by ABA, including the electricity supply. The company did subsidize electricity to the miners with the first 400 units of power being provided free. But the supply was often erratic and during the peak period there frequently was insufficient power for cooking. The countryside around the town was desert and there was always a great demand for wood, which made heating and cooking in the constant heat and dust all the more difficult. Despite these handicaps, many of the miners enjoyed a good social life, playing cricket, tennis and golf and attending the Saturday night dance. The golf course was popular, even though the greens were made from sand scrapes and the course itself was desert scrub. The Anglican Church also played a role in the community's social life by helping families which had been deserted, or where the husband had been sacked.

This was important because the company could, if it chose, refuse to allow a miner to leave town or to give him and his family support of any kind. Wittenoom was controlled completely by ABA, and it was not even possible to enter or to leave without the company's knowledge. That did not have any sinister connotations, but was an inevitable part of life in an isolated community.

Although the mill and the mine were eight kilometres from town, the families of miners were not immune from exposure to asbestos fibre. Tailings were used in all parts of Wittenoom as a surfacing and dressing material. They were laid about the race course, the parking areas, the school and the golf course and they were used around most domestic dwellings.[48] According to Dr McNulty of the Department of Public Health, ABA was advised against the use of tailings as early as 1958, but the practice continued unabated until long after the mine's closure in 1966. It is known that the aerodrome and the public parking areas were still being surfaced with tailings until the mid 1970s. Tailings were frequently used by the miners and their wives to keep down the red dust, and also because it seemed to encourage the growth of grass. Practices such as these help to explain the four recorded cases of mesothelioma which have occurred among Wittenoom residents who never entered the mine or the mill.

Most of the housing in the town was built and managed by the state Housing Commission which shows the extent to which the state government was committed to the interests of ABA. By March 1953, 152 houses had been erected by the state authority, but the company was still not satisfied that there was adequate accommodation, and it appealed for further construction work. In 1959 an agreement was reached between ABA and the Housing Commission that the company should take over responsibility for the management and maintenance of those properties which had, after all, been built especially for ABA employees. From that time the company had sole responsibility for all losses incurred through damage and unpaid rental, and it remitted a fixed annual sum in lieu of tenants paying rental directly to the commission. In the same year the rental of the houses was increased to around £3 per week with the money being paid to the company.[49] Father H. D. Middleton, the local Anglican priest, made a complaint about this arrangement to the Shire Clerk who subsequently wrote to the then minister for the North West, Mr C. W. Court.[50] Father Middleton expressed his disquiet about the new system whereby the Housing Commission used ABA as its agent, which resulted in a situation where the miners were living in "tied cottages". The arrangement most certainly gave the company the opportunity to increase its hold over the labour force. Not only did it control employment, transport, food, and basic services such as electricity and water, but it also now had a monopoly over housing. Any open dissent among the miners and their families was therefore impossible.

Australian Blue Asbestos was never satisfied with the amount of support it received from public authorities and, in 1960, it appealed for the provision of still further housing. At that time, there were 160 houses in the town, all of which were fully occupied and there was a waiting list for accommodation.[51] The company approached the minister for Police and requested that a new residence be built for the local constable, so that his existing dwelling could be released to the company for its own use. Increasing production at the mill was the main reason for the housing shortage, but it did not occur to the company that it should pay for housing its own workers: that had always been a matter for public expenditure. The commission was quite willing to comply with the ABA request, and ten extra houses were constructed in 1960 with a number of further dwellings being commissioned for the following year.[52] The company was, however, less diligent when it came to sanitation and the provision of an adequate water supply, both of which were paid for directly from ABA funds.

Each house at Wittenoom had the same aboveground water piping which, in the heat of the northwest, meant that there was no ready supply of drinkable cold water. Most of the water used for drinking and other uses came from a spring in Western Gorge. From the gorge a pipe brought the water to a reservoir from where it was piped to the town. Dennis Flowers, who worked in the general store, recalls that the water was often contaminated with asbestos fibre from the tailings which seeped into the reservoir.[53] The water also contained animal and vegetable matter and was unpleasant to drink. A letter from the Deputy Commissioner of Public Health, dated December 1957, proves that the situation was known to the public authorities. In the letter, Mr W. S. Davidson makes the following comment about the Wittenoom water supply: "The results of the bacteriological sampling of the water, provided the samples were properly taken and packed, indicates an unsatisfactory supply and all drinking water should be boiled."[54]

Although there is little further evidence regarding the water supply, Public Health department files do suggest that sanitation at Wittenoom was poor. Departmental inspectors visited the town in May 1956 and carried out a survey in the company of the ABA manager, Mr R. Jones. At the mine they found the staff mess and quarters to be in a filthy condition.[55] In some places the ceilings were falling down and the grease pits were filled with cockroaches. Conditions in the town had, however, improved since previous inspections and the flour in the bakery was now free from weavils, although the baker still smoked while working and the bread was sold unwrapped. The town butcher sold meat wrapped in newsprint, but the grocer had been able to overcome the problem of cockroaches. In the town cafeteria the inspectors found a dirty grease trap, broken plates and a septic tank which was not working. The cafeteria was supposed to service the needs of the single men, but it was found that many of the men preferred to prepare their own meals which they would eat in their rooms. This meant conditions were dirty and unhygienic throughout the single men's compound. Most of the buildings were overgrown with Chinese Lantern creeper which acted as a breeding ground for vermin. A number of huts were also badly damaged and, in the Italian quarter, the cistern was overflowing on to the ground. During the next eight years conditions in the town improved little.

A Public Health department inspection carried out in October 1965 on behalf of the Tableland Shire Council found sanitation was unsatisfactory.[56] The bakery building was unsealed and bread was still being sold unwrapped in violation of a local bylaw; the aerated-water factory, which was a small concern supplying local needs was judged unsuitable for food manufacture with the ingredients for the product being mixed by hand in a plastic bucket. An inspection of the ABA compound revealed unsanitary conditions, especially around the single men's compound. Behind the huts were heaps of empty beer cans and bottles together with scraps of food and other rubbish. The concrete wash-house and toilets were poorly constructed and unhygienic. The inspectors commented, "A revolting smell of stale soap and urine is noticeable in these buildings."[57] Refuse from the town was being disposed of in 44-gallon drums in a dry river bed on Wittenoom's outskirts and, when flooded, the rubbish was carried for miles along the creek.

The inspectors also found that much of the housing was unsatisfactory, despite the fact that it had been built at public expense and, supposedly, was being maintained by ABA. Many of the houses were timber-framed without interior lining which would leak during rain. The better dwellings featured external asbestos cement sheeting but, in general, they were poorly designed, particularly the bathrooms where tiles were decayed or absent and cisterns invariably leaked. The report also mentions the problem caused by the use of tailings throughout the township. It found this material used on driveways, footpaths and as a filler for lawns. The inspectors note, "It is considered that this is a danger to health and should not be permitted. The Public Health Department has contemplated legislation prohibiting and also compelling coverage of all existing tailings."[58] The department conveyed this opinion to the manager of ABA, Mr Lloyd, and he promised that in future the practice would be forbidden. Evidence from former residents, however, indicates that nothing was done and tailings continued to be used in this way for another ten years. The report was filed in 1965, that is, more than five years after the publication of Wagner's South African study on the threat of mesothelioma from environmental exposure. Australian Blue Asbestos was either unaware of the existence of Wagner's research or indifferent to its implications for the health of all those living at Wittenoom.

Responsibility for the monitoring and inspection of the mine and mill was shared by the departments of Mines and Public Health, each of which brought a different perspective and philosophy to its task. The Mines department had the role of making regular checks at the mine site and its inspectors filed annual reports. These men were experienced in the mining industry, but they also had responsibility for the mill which fell under the Mining Act and the Mines Regulation Act. It was an area in which the inspectors had far less experience. The role played by the Department of Mines was confused by its relationship with Public Health authorities, both state and federal, which have at various times played a part in the state's mining industry. From 1926 the Commonwealth took responsibility for the periodic examination of miners, but the state Department of Mines continued to be responsible for health and safety within the industry.[59]

Under the Acts a system of periodical medical checks was demanded, but hygiene practices within the industry were never coordinated with the medical data thereby gained. The process was quite unhinged and there was little communication between Commonwealth and state authorities. In 1953 the state Health

Department took over the monitoring of the miners' health, and from that time some effort was made to coordinate the procedure. And yet the same problem of fractured responsibility remained. Expertise about health and the judging of hygiene matters lay with the Public Health department, but authority for the regulation of the industry lay in the hands of the Mining department.

In the early years of the mine's operation, X-rays in the northwest of the state were carried out by mobile Commonwealth medical units stationed at Kalgoorlie. In the 1950s there was an annual X-ray programme at Wittenoom and after 1958, when the first case of asbestosis had been diagnosed, this was supplemented by thorough annual medical examinations.[60] When the mine first opened in 1944, there was no occupational health division within the Western Australian Department of Public Health, and the issue of workers' health was of little concern to public authorities in a state which boasted little secondary industry. In the case of Wittenoom, the isolation of the mine and the division of powers between the Health and Mines departments made regulation impossible. The Mines department had the powers of inspection but no expertise, and the Department of Public Health had the knowledge but no authority.

Public Health authorities first became aware of a danger at Wittenoom in November 1949.[61] A surviving minute from the Commissioner of Public Health, Linley Henzell, to the then minister for Health in February 1950 contains the following comment: "I was up at Wittenoom Gorge at the end of 1948 and found that the dust hazard in the present asbestos plant to be very serious. The whole of the surrounding area within a radius of one mile or more was smothered in the dust which is produced in the treatment. As far as was ascertainable very few, if any, precautions were taken. Asbestos dust when inhaled constitutes a very grave risk and is, if anything, worse than silicosis."[62] Health department officials did have some contact with their counterparts in the Mines department who, when asked, made available dust readings carried out at Wittenoom. In 1951 the permissible level of dust in Western Australia was set at 176p/cc. A letter from the Commissioner for Public Health to the Under

Secretary of Mines in October 1951 refers to the high readings found at the Wittenoom mine.[63] The Commissioner expressed his concern at the situation and then went on to comment on the danger suggested by the readings: "Asbestos produces more rapid fibrosis of the lung than silicosis and is more liable to superimpose infection. It is therefore apparent on the data submitted than conditions are far from satisfactory and in the interests of the health of the workers, this department must investigate the matter further and press for immediate improvement of supervision and working conditions at Wittenoom."[64]

The concern of officers from the Department of Public Health is apparent in departmental files throughout the next ten years. In the annual report for the year 1959, there is a discussion of the asbestos industry and of the existing dangers to the health of mine and mill-workers. The report notes that there had been no new cases of asbestosis during the preceding twelve months, but that past experience shows that the length of exposure for men contracting asbestosis or silicosis is far shorter than for men engaged in the goldmining industry. This fact was acknowledged to be disguised by the high labour turnover and by the small numbers of men involved in asbestos mining.[65] In the period until 1958 there had been six cases of asbestosis recorded at Wittenoom, of whom five were mill-hands who had an average period of exposure of only four years. Each of these men had been advised to leave the industry but all had decided to stay on. The department viewed the situation at Wittenoom as presenting a "grave hazard".[66] The author of the annual report for 1959, Dr McNulty, discusses the nature of the problem which demanded, in his view, a close liaison between medical and engineering specialists. He concludes his report with the following comment, which must have been brought to the attention of CSR and the manager of ABA: "Despite the many marked improvements which have been effected at the Mill and the Mine, I am not satisfied that the risk of industrial disease has been eradicated or even brought to par with the risk of silicosis in the gold mining industry."[67]

Apart from the organizational confusions it shared with the Mines department, the Department of Public Health was handicapped by a lack of data about occupational health. Before the war there had been no system of notification of occupationally related disease and injury and, even in the state's major hospitals, a patient's occupational background was rarely scrutinized as a possible cause of his or her illness. Only in 1952 was a system of notification and the recording of data introduced.[68] The Public Health department was well aware of the problem and, in 1953, a delegation from the department approached the minister, the Hon. El Nulsen, in the hope of achieving some change to the existing legislation.[69] At that meeting, Dr Henzell argued against the existing system, which saw his officers prevented from exerting any influence on matters of industrial health when they fell under the jurisdiction of other acts, such as the Mines Act. He went on to comment, "There is an industry where men are working underground, and the regulations concerning sanitation, etc., may be made without any reference to the Commission who has expertise on such matters."[70] Dr Henzell noted that he had approached the Mines department to permit his officers access to those areas of health which came under the Mines Act. His approach had met with rebuff. After listening sympathetically, the minister agreed that a change was justified but no change resulted. The proposals suggested by Dr Henzell would have required the amendment of numerous acts, including the Inspection of Machinery Act 1921–1951, the Health Act 1911–1952, the Mines Regulation Act 1946, the Road District Act 1919–1931, the Municipal Corporations Act 1906–1951 and the Town Planning and Development Act 1928–1949.

Whatever its limitations as a body regulating industrial hygiene, the Department of Mines did manage to produce a quantity of documentation about conditions at Wittenoom. In an inspection carried out in 1945, officers from the department found that a serious problem with ventilation existed and that dust levels had increased over those found in previous years.[71] The inspector believed, however, that the problem was due to wartime conditions and also to the isolation of the site which created difficulties with transport, manpower and plant. The Mines department itself suffered from a lack of trained officers and transport, which made the regulation of outlying mines difficult.[72] Obviously, the inspectors felt some empathy with ABA in its struggle to establish a mine in one of the most remote parts of the state. In 1948 two major inspections were carried out by Mr Lloyd, the state Mining Engineer. In his first report, filed in February, Lloyd mentions the dust readings as being consistently high throughout the mill, averaging around 300 p/cc.[73] At the time of Lloyd's visit, the mill was not in full operation, and no dust counts were taken in the mess room. Even so, Lloyd was led to comment: "However, there was sufficient evidence of dust in the building to warrant a request that after one week the new dining room, situated some three-quarters of a mile from the mine, should be made available."[74] He informed the company of his views and he was given a guarantee that action would be taken. At the time of the inspection Lloyd found a total of one hundred and twenty employees of whom seventy were engaged in the mill and fifty underground.

In his second visit to Wittenoom in 1948, Lloyd was pleased to discover a marked improvement in the mine's ventilation.[75] But there was no such improvement in the mill where the conditions " . . . from a dust point of view were found to be most unsatisfactory."[76] Five dust samples were taken in the mill and three of these showed readings of above 1,000 p/cc. Lloyd drew the attention of the manager to the dust and to what he perceived as a "danger to the health of the men".[77] The manager told him that the existing plant was to be discarded and that work would commence on a new mill as soon as sufficient data had been collected on the existing mill, so that any flaws in design would be avoided in the new building. The dust readings taken by Lloyd were subsequently forwarded to the manager of ABA.

In the annual report for 1948, the year of Lloyd's inspections, the department appears to have been satisfied with the efforts being made by ABA, a company working under difficult conditions. The report notes the general effort towards improvement and that ventilation in the mine was satisfactory, although this was certainly not the case in the mill.[78] The department accepted ABA's explanation that a new mill was to be built and that the old plant, although unhygienic, would be used only until such time as it could be replaced. In fact, the opening of the new plant still lay some ten years into the future, and the Department of Mines made no effort to obtain a concrete proposal from ABA about its construction. In 1949 the mine had begun to increase output and the department's annual report mentions the need for dust-monitoring equipment and ventilation, but there is no suggestion that it was dissatisfied with the efforts being made by ABA. The annual report for the following year does mention the department's concern with the question of ventilation throughout the asbestos industry. It also suggests that mine owners were making every effort to reduce dust levels where necessary: "The majority of the mining companies are cooperating to the fullest extent with the department in attacking all ventilation problems."[79]

The report also refers to the fact that the District Inspector of Mines, Mr J. E. Lloyd, attended the International Conference on Pneumoconiosis held in Sydney during the year. It is curious that the department, with its limited funds, should take the trouble to keep abreast of developments in the medical field concerning pneumoconiosis, and yet fail to ensure that the promises made by ABA in 1948 about the construction of a new mill were kept. The subject of the new mill disappears completely from the annual reports for the next eight years.

There is little indication in the annual reports of a chronic problem of exposure in the asbestos industry, even though there is ample evidence from the department's own files that conditions at Wittenoom, the state's major asbestos mine, were hazardous. Documents dating from 1952 mention the dust hazard in the mill,[80] and the department instructed ABA to enclose the area concerned and to install larger extractor cyclones. Apparently, some work was completed, but production levels were also increased so that the same problems soon returned. In 1952 ABA produced asbestos fibre valued at £200,000. In the following year this figure had more than doubled to £558,000. In 1953, the number of men employed at Wittenoom rose to more than two hundred and output was increased substantially. With the successive increases in production, the old mill was placed under greater strain so that the problem which Allan Osborne had noted in 1950 became acute. Surprisingly, this burden on the existing plant went unnoticed in the department's annual reports which, from 1956 until the mine's closure in 1966, contain no criticism of ABA or of its methods of operation. In 1961 the annual report says: "The company [ABA] is making every effort to effectively control dust in both mine and the mill."[81] The same shallow optimism is found in other years and often there is strong support for a company which the department always viewed as having to work under great difficulties. The delay in the construction of the new mill dragged on for so many years that along with the unfulfilled promise of a "wet process", which in the annual report for 1956 was mooted by the department on the company's behalf, it now appears fanciful. The mine's closure in 1966 was explained as being due entirely to economic factors, as successive years of a trading loss had eventually forced the company to withdraw. It was certainly not forced upon ABA by the Department of Mines demanding that it comply with existing hygiene legislation.

Mr John Faichney, who was employed by the Mines department as an engineer, visited Wittenoom regularly in the period from 1951 until 1958. It was his job to take dust samples in both the mine and the mill, although most of his attention was paid to the underground workings. He recalls that dust levels in 1951 were set at 300 p/cc, but he was more concerned with the threat of silicosis than with asbestos fibre.[82] He now admits that he had no idea about overseas legislation relating to asbestos dust, or even of the existence of the disease of asbestosis. The department did not take any specific action against ABA over the frequently high counts, but Faichney did recall that from 1950 onwards miners were subject every two years to compulsory chest X-rays.

Evidence from former miners suggests that inspections at Wittenoom were in any case inadequate, because of the deliberate slowing down of the mill which reduced the dust levels whenever inspectors were present.[83] The company was always aware of pending visits by Mines department officers, simply because it was impossible to arrive unheralded at such an isolated town. There was only one road into Wittenoom from Roebourn and the arrival of any visitor was always the subject of interest.

Throughout the years of the mine's operation ABA had diffi-

culty in retaining the services of a physician. At one time the
company sought to have acceptance of overseas qualifications
by the AMA, because it was impossible to attract a qualified
doctor. In 1953 ABA approached the minster for the North West
about the need for a resident doctor to service the seven hundred
people living in the town.[84] The absence of professional medical
care left the community and, in particular, the miners in a vulnerable position. Mining has always been a hazardous occupation and accidents are common. The town was visited regularly
by the Flying Doctor Service based at Port Hedland, but this service was inadequate both for day-to-day needs and in the case of
serious accident. The company advertised throughout Australia
but failed to gain a single applicant. This problem persisted
throughout the 1950s and the reason may be seen in the correspondence of Dr Gordon Oxer who worked at Wittenoom during
the early 1960s. The Department of Public Health at that time
questioned the level of fees being charged by Dr Oxer, which he
admitted were higher than those allowable in Perth. In a letter to
the department, Oxer explained that his fees were fully justified
because of the conditions in which he was forced to work.[85] He
lived in virtual isolation from his own family who had remained
in Perth, and from his professional colleagues. He lived in
primitive conditions, and as the only qualified physician in an
area of some thousands of square miles he was faced with a high
degree of responsibility. Oxer requested the right to further increase his fees and it is obvious from his letter than he was unhappy with his lot and with the conditions of his employment. In
fact, his position as a private practitioner and his relationship
with the company were highly ambiguous. According to departmental files,[86] ABA regarded Oxer as its employee. His salary
was in part guaranteed in an agreement between ABA and the
Department of Public Health which gave him an assured income
of £3,700 per annum, with ABA and the department reimburs-
ing him for any amount below that figure. Dr Oxer's independence was further eroded by the practice in which ABA collected
fees directly from his patients, presumably with the costs being
deducted from the miner's wages. Under such circumstances it
would have been difficult, if not impossible, for the town physi-

cian to question the company's hygiene practices. The absence of an independent voice about the way in which the mine was being run made the role of the trade unions at Wittenoom all the more important.

Most of the workforce was covered by the Australian Workers Union (AWU), while a few skilled tradesmen came under the protection of the Amalgamated Engineering Union (AEU). There was little trade union activity or influence at Wittenoom. The AWU was quiescent, and the AEU tended to discourage its members from working for ABA.[87] The reason had nothing to do with industrial hygiene, for like the AWU the AEU was concerned more about wages than industrial illness or injury. For the miners, the right to strike did not exist. They lived in houses constructed for the company's benefit with rental being collected by company officers, and all power and water were provided by ABA, which also ran the only store where food and other basic supplies could be bought. Most of the miners were indebted to ABA and they could not leave the town without permission of the company which controlled virtually all transport. Any workman who objected to the conditions of employment was faced with immediate dismissal and the loss of housing. In an isolated northwestern town the consequences of opposing ABA would have been catastrophic, both for the miner and for his family. The degree of control exerted over the community by ABA, and the composition of the workforce, its ethnic diversity and the short-term nature of most employment, each worked to divide the community. Miners had little sense of solidarity with their fellow workers and the language barrier meant that it was common for men to be unable to communicate, except at the most basic level. There was little overt hostility between the different ethnic groups, but national loyalties tended to fragment an already diverse labour force. Despite these barriers, there were occasions when workers did join together informally to direct the labour process to their own ends.[88] At various times the work foremen would try to push the men to increase their production quota. These demands made conditions in the mine even more difficult, and also resulted in the bagging of an inferior product as the fibre would contain much gravel and stone. In protest, the workers would place metal pieces on the conveyor belt which, when it entered the crusher, would grind the machinery to an immediate halt. This was something of a standard practice whenever the system was worked too fast, and it gave the men a welcome break.

Although there was little trade union activity at Wittenoom from as early as 1948, the AWU was aware of conditions in the mine. In that year Charlie Oliver, an AWU official, approached the Mining Engineer at the Mines department about what he believed was a health hazard to the miners from asbestos dust.[89] Oliver claimed that, when the mill was operating, heavy clouds of dust would drift throughout the plant into the engine room, the fitting shop and even into the mine offices. He was in no doubt that the workers' health was at risk. Oliver's complaint resulted in a successful wage claim, which also held the promise that conditions in the mill would be made safe. In 1950 the AWU succeeded in its case before the Arbitration Court for the payment of various benefits at the mine. Under this award,[90] all workers engaged by ABA in Perth or elsewhere within the Commonwealth were to be provided with free transport to Wittenoom with the amount to be deducted from their wages. After six months' continuous employment this sum was to be refunded.

The award makes specific reference to the dry crushing plant, one of the dirtiest parts of the mill. Under the award, this plant was not to be operated unless the dust levels were reduced "as far as is reasonably practicable". To ensure that the company complied with the award, the Mines Inspectorate was assigned to visit the mine each month, or when requested by the AWU. The award also specified the need for water jets to reduce dust and for proper ventilation, a subject which is covered in some detail. In the event of a dispute over ventilation between the union and the company, the award allowed for the setting up of a Mines Ventilation Board, a tripartite body with representatives from the trade unions, the ABA and the Mines department. Despite all these statutory precautions against dust, the award also introduced an above-award allowance for men exposed to high dust levels.

The right of workers to receive this allowance was re-negotia-

ted in 1953 and 1956, when the Board decided to introduce an
allowance per hour for all those employed in the mill and its surrounding. The Board accepted the dust as an unavoidable part
of the asbestos industry about which ABA could be expected to
do nothing. In 1957 the AWU went to the Board once more to
have the allowance increased from one shilling and sixpence to
two shillings per hour for all men in the mill and workshop.[91]
The Board's decision was handed down in October 1957. The
Board found that there had been a marked improvement in the
dust extraction system at Wittenoom, and it rejected the AWU
claim and reduced the existing allowance. The new allowance
was set at fourpence per hour for workers in the remaining dusty
areas and threepence per hour for men working in the bagging
room. Not all those people familiar with conditions at Wittenoom shared the Board's opinion that conditions had improved
and there there was no risk to the health of ABA employees. The
Board's judgment was certainly not shared by inspectors from
the Department of Health who visited the mine.

One of the key figures in the history of the Wittenoom mine is
Dr James McNulty, who diagnosed the first recorded case of
mesothelioma caused by asbestos in Australia. That case was
diagnosed in 1961 in a man named Joseph Sawer who had worked at Wittenoom from 1947 to 1949. Soon after the diagnosis
Sawer had died, but it strengthened McNulty's concern about
the mine and the dangers to the people living there. McNulty
first visited Wittenoom as a health inspector in 1958, and he was
alarmed at what he found. According to McNulty, the Health
department issued a warning to ABA in 1948 about the dust and
the possible effect on health, but nothing was done.[92] The working conditions at the mine like those in the town itself were poor,
but it was not until 1958 that the first case of asbestosis was discovered in a miner also suffering from tuberculosis. At about
that same time, routine chest X-rays began to reveal signs of
pneumoconiosis. This meant that, despite the fears of Health
department physicians, it had taken fifteen years for the proof
of a health hazard to become obvious. When he first visited the
mine in 1958, McNulty was disturbed not only by the work conditions but also by the use of tailings from the mine. The

material was spread everywhere: on the golf course, on the race course, and around the houses as a form of top-dressing. McNulty recalls, "I passed the remark then that it seemed a shame that they are exposed to it at home."[93] McNulty's observation was made before the publication of Wagner's study implicating asbestos in the occurrence of mesothelioma through the medium of environmental exposure. Although McNulty could find various reasons why ABA could not control the dust in the plant, at least part of the problem was due, so he believed, to the company's lack of expertise. Australian Blue Asbestos tended to employ senior personnel drawn directly from the sugar industry, rather than to recruit from outside the parent company, CSR. These men had no experience in running a mine and they would have seen few career prospects in becoming expert in the asbestos division of CSR.[94]

Throughout his years of visiting Wittenoom, McNulty felt a strong sense of frustration. Like other officers from the Public Health department he could do nothing to improve the work conditions, even though he and other physicians who carried out X-rays on miners were in the best position to judge the dangers to their health. When the mine closed in 1966, McNulty admits to being near "rebellion" about the way the mine was being run. The division of powers between government departments made the situation of doctors such as McNulty very difficult. He recalls of his days at Wittenoom, "As doctors we were faced with a serious moral dilemma in giving the men the clean health certificates they needed to work in the mine."[95] The physicians were aware that, although there was no immediate evidence of illness, pneumonocosis was inescapable, and yet the men demanded the right to work. Despite the repeated warnings from the department and from concerned physicians such as McNulty, this dilemma was not felt by the managers and owners of ABA.

5

Wittenoom: The People

When CSR entered the building materials industry in 1938, it became interested in the manufacture of asbestos cement sheeting. The company had originally planned to fabricate building materials from the by-products of sugarcane refining, and a small plant was established for that purpose. But CSR did not become a building materials company, and its management never came to terms with the demands of an unfamiliar industry.

Australian Blue Asbestos recorded losses for most of the years of its operation and, with the exception of a brief period from 1956 to 1961, the company was unable to operate successfully. By 1955 ABA had accumulated loses of £800,000 and at the closure of the mine in 1966 this figure had grown to $2.5 million.[1] Australian Blue Asbestos was never a part of the parent company's long-term strategy, and its failure to return a profit merely discouraged CSR from further investment, or in seeking to reverse its poor performance. Investment at Wittenoom may have been appealing in 1944 at a time when asbestos was a strategic material. But in the post-war years CSR was not able to succeed in fabricating asbestos products successfully in competition with James Hardie, or in producing fibre from the mine at Wittenoom competitively against imports from Canada and South Africa. Whatever the suffering that the Wittenoom mine has brought to the lives of the former miners and their families, that suffering was not translated into profits for the parent company, CSR.

When the question of illness among ex-miners became a matter of public controversy in Western Australia in 1978, Lang Hancock immediately sprang to the defence of the industry.[2] Hancock, who had repurchased the mine and many of the tangible assets at Wittenoom in 1966, claimed that in the case of asbestos, as with so many of the products of modern industry, some people had to suffer so that many could benefit. He argued that asbestos in products such as brake shoe linings had saved countless lives whereas only a handful of miners had died from inhaling fibre. Although he expressed some sympathy for the men and their families and even though he had shared ownership of ABA during the first years of its operation, Hancock declined to become involved in any compensation scheme for the victims. His comments outraged many people and he was taken to task in the daily press for his insensitivity. Hancock's critics objected to the equation he drew between the economic benefits which flowed from the asbestos industry and the ruined health of a few miners. And yet this is exactly the line of argument which the Australian asbestos industry had used for some years without attracting the least public censure.

When writing in the *Medical Journal of Australia* in 1974, Dr S. F. McCullagh, a senior medical officer with James Hardie, argued a strong case against the supposed dangers from environmental exposure. Dr McCullagh claimed that asbestos has always been present in the general environment and in human water supplies, and that there is no hazard from airborne fibre released from brake linings. He went on to argue in much the same way as Lang Hancock would four years later: "The industry is well aware of the hazards of asbestos, and having briefly reviewed these I think we should also remember that, if we consider no more than its fire-retardant properties and its use in brake linings, asbestos has saved more lives than it has claimed. With the great improvement of standards of industrial hygiene over the past decade, this credit balance, if I may so call it, will increasingly grow more favourable."[3] Like Lang Hancock, Dr McCullagh had no difficulty in drawing up a balance sheet between human suffering and the presumed economic advantages to the community as a whole brought by the asbestos industry. Unlike Hancock, he declined to draw attention to the issue of the high profits which certain sectors of the Australian industry had enjoyed in the post-war years. It is not possible, however, to judge the morality of the industry without taking account of the actual human costs which asbestos-related diseases bring to victims and their families.

When Nick D'Ascanio first joined ABA he could speak no English, and his interview for employment was carried out through the medium of an interpreter.[4] He remembers being given a chest X-ray at a local clinic in Perth before being sent to Wittenoom. His first job was as a cleaner, working directly beneath the conveyor belts which carried the ore into the mill. He was provided with a face mask as protection, but this proved useless as the single pad provided for each day's work would soon clog up, and he was unable to breathe freely. In his second week, he was moved to the bagging station. It was very hot and he did not wear a shirt so that the fibre clung to his skin. He would hose the fibre off his arms and legs during the regular fifteen-minute breaks, but he was always uncomfortable because of the airborne dust which was everywhere. D'Ascanio recalls: "When we worked the night shift, the next morning while we were waiting for our transport to take us back to our accommodation, through the rays of the early morning sun the dust was clearly visible — it was all over the place. The asbestos dust covered a radius of hundreds of metres."[5] After spending the first four or five months at the mill, he was moved into the mine. It was his job to shovel into skips the ore which could not be reached by the hydraulic shovel. The material was very dry and dust was everywhere. The ventilators, which were supposed to clean the air, managed only to circulate the fibre so that even when the ore face was not being worked, dust was visible in the miners' headlamps. Twenty years later D'Ascanio is now an invalid pensioner suffering from asbestosis. His family's weekly income is $120, on which he has to support his wife and two children. Asbestosis is an illness which makes most ordinary tasks unpleasant or impossible. It is a disease that gets progressively worse and for which there is no effective treatment. In severe cases the patient will end his or her days breathing through a respirator in a hospital bed, unable to walk freely and quite incapable of participating in an ordinary family life. It is a completely debilitating disease, as distressing for the patient as it is for his or her family.

Oscar Penetta arrived in Australia in January 1957 and began work with the Perth Water Supply.[6] After five months he took a job as a machine miner with ABA. He was attracted to Wittenoom by the promise of high wages, but the heat and discomfort were such that he stayed initially for only a few months. Penetta soon returned bringing his wife and children. He was involved in blasting ore away from the pit face. This was done three times each week and the miners would re-enter the shaft before the dust had settled. Sometimes the AWU representative would stop the men from going back to work until the air was clear, but it made little difference as there was always dust in the tunnels. Over the next five years Penetta went to and from Wittenoom on various occasions drawn back each time by the high wages which he could earn with overtime. In 1969 he began to suffer from chest pains. He became concerned but his X-rays at the Perth Chest Clinic showed no evidence of disease. The pain persisted and Penetta ws re-examined by his own doctor who diagnosed a 40 per cent disability with asbestosis and silicosis. Over the past seven years Penetta's condition has worsened and he now has a 75 per cent disability rating. Penetta, like D'Ascanio, is faced with irreversible decline in health and he sees little prospect of being able to lead an enjoyable life. When first diagnosed as having asbestosis, he was terribly upset and he is now forced to live with the knowledge that his illness will become progressively worse.

Arthur Ballerum was born in Sydney and went to sea as a young man.[7] He worked on the Fremantle wharves where he handled asbestos bags from the Wittenoom mine. He was employed later on various vessels working the northwest coast ports, including Port Sampson which was the major handling centre for Wittenoom asbestos. The fibre was packed in hessian bags, and over a four-year period Ballerum called there weekly. During the 1960s Ballerum left Western Australia and wandered from job to job mostly in New South Wales. After several years of declining health, Ballerum had a chest X-ray, which revealed asbestosis, and he was awarded a 50 per cent disability rating by the Workers Assistance Commission. He cannot remember ever being warned about the dangers of working with asbestos, but he does recall that many of his old workmates from the wharves have, in middle age, died from respiratory diseases. Ballerum now has an uncertain future blighted by chronic illness. He cannot trace the insurance company liable for workers' compensation for the state-owned ships on which he worked and where he contracted his disease.

There are numerous cases like those of Penetta and Ballerum and most contain many of the same elements. It is possible to list such cases one after the other without achieving any greater understanding of the effects of asbestos-related disease or of the human costs which accompany chronic illness. There is one case, however, that of Mrs Joan Joosten, which does demonstrate the myriad of ways in which asbestos intrudes into every aspect of the life of the patient and his or her family. The Joosten case is important, too, because it was something of a test of the rights of workers to sue their former employer, ABA. Mrs Joosten was not a miner, but she was employed in the ABA company office at Wittenoom for a three-year period during the 1950s. It was more than twenty-five years later that she was first diagnosed as suffering from mesothelioma. Her common law action was brought against Midalco, the surviving subsidiary of ABA, and it was funded by herself and her husband with the help of two sympathetic lawyers. Joan Joosten's reason for bringing the action was to benefit other sufferers and not herself, for she knew that she was unlikely to live to see a favourable verdict. If she had succeeded in her claim for damages the rewards would have been small, perhaps as little as $50,000 as compensation for her lost earnings as a stenographer, Joan Joosten died on 10 March 1980 less than one hour before her appeal to the Full Bench against the dismissal of her original claim was to be heard.

Joan Joosten was born in Papua New Guinea in 1927 and at the age of five she moved with her family to Perth. Her mother was a teacher with the Education department and the family moved frequently from one small rural town to another. Joan was a good student and she passed her leaving certificate which allowed her to secure a job with a Perth bank, where she was employed as a bookkeeper and stenographer. As a young woman she was very active, enjoying walking and swimming. She married Hubert Joosten in 1950 and, after a brief honeymoon, she and her husband moved to Wittenoom where he was employed as a fitter and milling machinist. They were attracted to Wittenoom by the high wages and by the prospect of being able to get a start in life. Hubert Joosten was employed, initially, in the workshop where he later became foreman in charge of all repairs to the mill. As most of this work had to be done on the weekends so as not to halt production, there was ample overtime which gave Joosten a high wage, far higher than he could hope to earn in Perth.

Hubert Joosten remembers the mill as being extremely dirty. During some periods the mill and the mine were worked continuously in three shifts, and the machinery and the dust-extracting equipment were constantly breaking down. In the mill itself the electric lights were kept burning all day and all night. Joosten recalls, "On the rafters they had, if I remember, light globes of 250 watts, and when it was a quiet day with not much wind and it was working full blast, the lights of those 250 watts were just like candles, there was that much dust in the mill."[8] Hubert and Joan Joosten lived in the town in a timber-framed asbestos sheeting house. It had a corrugated iron roof with two bedrooms, a kitchen and a dining room. The house was located near the town centre, directly opposite the tennis courts. Joan worked in the company office which was situated less than one kilometre from the mill and, with their combined wages, she and Hubert were easily able to save. Joan enjoyed walking, rock climbing and tennis, but she had little free time after her work as a stenographer and after looking after the home. She would leave for work at 8.00 each morning in the company bus for the mine which was eleven kilometres away and return at 4.30 each afternoon.

In evidence at the trial, Joan Joosten recalled that it was always obvious when the mill was operating because of the dust spewing from the chimneys on the millhouse roof. On a still day a cloud would hang above the Gorge so that, even in the company office, there was no escaping the dust and fibre. The flywire screens surrounding the office were always clogged with asbestos fibre

Wittenoom

(All photographs supplied by Asbestos Diseases Society Inc., WA)

1. The Wittenoom mine was built on three levels down the side of a steep gorge. The crude ore was fed into the mill at the top with the processing taking place at various stages until the fibre had been extracted at the lowest level. At its peak, the mine and mill employed more than 400 men, most of whom were recruited in southern Europe to work for Australian Blue Asbestos.

2. A section of the housing provided by ABA for its employees at Wittenoom. Conditions were sparse and because of the heat and the isolation, living conditions were difficult. The high wages paid by ABA continued to attract labour to the mine.

3. Tailings from the mine were used to top-dress roads and paths about the miners' homes. The tailings were also used at the airport, on the golf course and at the school.

4. The fibre from the ABA mill was packed by hand into hessian bags and shipped by truck to the coast. Although plastic bags were available, hessian was preferred by the company even though the loose weave allowed fibre to escape.

5. A willi-willi approaching a miner's home at Wittenoom. There was always a shortage of water in the town and dust was a constant problem. The use of asbestos-rich tailings from the mine meant that the dust invariably contained some asbestos fibre.

6. The Wittenoom airport, c. 1955. The mining company ABA controlled all means of transport in and out of the town and it was impossible for a miner and his family to leave without the company's consent.

7. The entrance to the Wittenoom mine, *c.* 1958. The pipe leading into the shaft at the top right conveyed compressed air which was used for drilling.

8. Working the mine face.

9. A group of miners, *c.* 1960. There is no record of any Aboriginal labour being employed by ABA and it was company policy not to employ blacks.

10. A miner operating a scraper in the Wittenoom mine. Conditions in all parts of the mine were cramped, which made exposure to dust and fibre an inescapable part of daily work.

11 (a) and (b). Conditions inside the Wittenoom mine were cramped. The ore occurred in narrow seams and the host rock was extremely hard. This meant that the miners laboured in low shafts where dust and airborne fibre generated by the mining was inescapable. At no stage did ABA manage to effectively control the creation of dust.

12. A group of miners in the crib room during a break. Conditions in the mine were unpleasant. There was noise, heat and the constant problem of dust.

13. A funeral of an unknown person at Wittenoom.

which, like snowflakes, would become caught in the wire.[9] Her husband remembered that the dust in the office was so thick that it was possible to write one's name on the desktops.[10] There were no warning signs about the dangers from inhaling the dust and, during the three years she worked for ABA, Joan Joosten was never given a chest X-ray.

After leaving Wittenoom, Hubert Joosten had various jobs including periods spent working as a newsagent, and running a fruit shop and a sandwich bar. Joan worked for a brief time in the newsagency, but she spent most of her time doing housework and gardening. As always she led an active life and she would run each day along the foreshore near their home in the Perth suburb of Como. In 1978, when local newspapers began featuring stories of illness among ABA miners, Joan became concerned about the health of her husband and she persuaded him to have a chest X-ray. The result was negative but soon after she herself became ill. Following heavy work of any kind she would become suddenly tired and she experienced constant pain in her arm. She also began to lose weight, and in less than three months went from being a robust woman of 76 kilograms to a frail 50 kilograms. But she did not seek medical attention until after viewing a television documentary about Wittenoom and the effects of asbestos. After learning that she had lived at Wittenoom, her physician suggested she have an X-ray. The X-ray showed that she had mesothelioma. The pain continued unabated and it made her very tired and short-tempered. She continued to lose weight and she was no longer able to lead a normal life. She could not work in the garden, which she had loved, and she could not do heavy work such as cleaning. After an hour of even light housework she would have to spend the rest of the day in bed, leaving the running of the home to her husband. The pain in her arm increased to the extent that she could no longer use her right hand. This made it difficult for her to comb her hair or to get dressed. The only clothes she could wear had to be buttoned up at the front, and even then it was a struggle to dress or to shower. Her breathing became less and less certain, and X-rays showed the tumour in her right lung had grown to the size of a tennis ball.

Because of the pain, Joan Joosten was unable to sleep and she spent most of her days lying on a couch to ease the pain. Hubert retired from work to look after her, and did all the cooking and housework; each afternoon he would take her for a drive. She never complained or spoke about the pain and she refused all medication, preferring to adjust her diet and to accept the discomfort, rather than to be drugged and incoherent. Even though he had sold his sandwich bar only so as to be able to nurse his wife, Hubert Joosten was ruled ineligible for any social security benefits when Joan died.

Justice Wallace handed down his findings in the Joosten case on 9 October 1979. He dismissed her claim and set aside the question of costs. The reasons for Wallace's decision reveal much about the kinds of problems confronting plaintiffs in asbestos-related litigation. There was no doubt in the court's mind that Joan Joosten's illness had been caused by her employment at Wittenoom. Mesothelioma is a rare disease which is closely related to exposure to crocidolite. The major issue facing the court was the question of foreseeability: should Midalco (ABA) have reasonably foreseen at the time of Joosten's employment that only a brief period of exposure, not in the mine or mill but in the company's office, would result in serious illness or injury. According to Justice Wallace, apart from the question of damages, this was the only issue at stake. Joan Joosten claimed that ABA failed to warn her or to take adequate steps to protect against airborne fibre. She claimed that the company should have located the office in which she worked away from the dust pouring out of the mill. However, Joosten's counsel acknowledged that the company could not have been aware at the time of her employment of the threat of mesothelioma. From 1950 to 1953 there was no medical literature connecting exposure to blue asbestos with her particular form of cancer. Midalco, the defendant, denied that Joosten's illness was due to exposure to asbestos while she was employed by ABA, and it also sought to have the case dismissed on the basis of the statute of limitations, which demands that an action be brought within six years of an injury. The judge was then left to decide the question of what steps the company should have taken to guard its employee against an unknown danger; of course, it could have taken none, and therefore the case was dismissed.

If Joosten's claim had been successful, it is unlikely that she would have gained financially from the decision. The case was fought against Midalco and not the parent company, CSR. At the time of the trial Midalco had paid-up capital of a little over $300, although it could call upon sufficient funds to fight the case with the help of senior counsel. It was not sufficient for Joan Joosten to gain the support of the court in its verdict: she would have had to gain a settlement against the parent company, which remained carefully out of sight during the proceedings. Even though she was unsuccessful, her case did gain national publicity for the plight of asbestos victims, and it unearthed a considerable amount of damaging evidence about the way in which the mine at Wittenoom had been operated.

Although most of the victims of asbestos-related disease are to be found among the ranks of former miners, there are four documented cases of mesothelioma among women who lived in the town and who never entered the mine. The most distressing case is that of a woman who died at the age of 28 after having been exposed to the asbestos-rich tailings about her home. As a child she, like many Wittenoom children, played in the tailings which were spread freely about the town. Another case is that of a woman who, during the 1950s, lived at Wittenoom. Her husband was a truck driver employed in carting asbestos bags and she worked in the local grocery store. It is probable that she contracted the disease after exposure either through washing her husband's work clothes or from the fibre found on the clothing of the customers who came into the store. Each of these women was a victim of the contamination which the ABA mine brought to the entire community. The right in law of these women was limited by the fact that their illness was contracted not under the terms of their employment but came from their place of residence. Neither they nor their families had any direct claim against ABA.

The nature of asbestos-related disease is such that the effects of the illness fall as much upon the families of the men and women who have contracted asbestosis, lung cancer and mesothelioma as they do upon the patients themselves. In most instances the quality of life of the entire family is devastated. This ever-widening circle of misery is not taken into account in the awarding of worker's compensation payments, or in the settlement of civil actions. It is, however, a reality well known to the families of ex-miners. Patricia Lambert was born in England and arrived in Australia with her husband and their children in 1963.[11] Her husband, Harry, was then 35 years old and was a trained boatbuilder. He was fit and actively involved in the Anglican Church. Harry Lambert began work in the northwest in towns such as Port Hedland and Tom Price. Much of his work involved sawing asbestos cement sheeting. The sheets were often cut in confined spaces with an abrasive disk fitted to an electric saw.

In 1979, Lambert became impotent and withdrawn, but a medical examination showed he was still in good health physically. During the next two years Patricia Lambert was aware that her husband was unwell, and in July 1981 she persuaded him to go to a local GP. In that same month Harry Lambert was diagnosed as suffering from mesothelioma and, in September 1981, he was granted an invalid pension. The diagnosis and Lambert's ever-worsening health changed the family's situation dramatically. Lambert could no longer work and the family was forced to sell its only assets to pay off existing hire-purchase debts. Like most people, the Lamberts lived in a web of long-term indebtedness and the loss of income destroyed the little security they had. Lambert at this time became irrational, and underwent what his wife would later describe as a "severe personality change". Every two weeks he would attend hospital to have fluid drawn from his lungs. The treatment was very painful and Lambert found the visits distressing. He also had a brief period of radium treatment, but the sessions were too painful to be continued. Often, after returning from the hospital, Harry Lambert would not know where he was.

By August 1982 he required round-the-clock nursing, and he entered hospital for several brief periods to allow his wife the opportunity to rest. Lambert died at home in December 1983. After approaching a lawyer over the question of worker's compensation, Patricia Lambert was told that her husband may have contracted the disease "anywhere". The only money Patricia Lambert had received up to the end of 1984 was the $12,000 payable under her husband's superannuation scheme and she is now living on a widow's pension of $181 a fortnight. Her initial approach to a local solicitor proved frustrating, as he failed to make any representations on her behalf and the case was effectively delayed for over twelve months. Like many people in her position, Patricia Lambert feels intimidated by professionals, such as physicians and lawyers, and she has difficulty speaking up for herself and in gaining her rights.

Harry Lambert was 56 years old at the time of his death and his wife was 54. In August 1983 the Pneumonocosis Board granted him a 100 per cent certification for mesothelioma, but this did not help to hasten the settlement of his widow's rights to a worker's compensation payment. The main obstacle faced by Patricia Lambert has been to identify the responsible party, as it is impossible to know where and when her husband contracted his disease. The continued delays have only worsened the strain on his widow, who is still recovering from the experience of her husband's illness and death.

The experience of Julia Armstrong was equally difficult.[12] Her husband, Ronald, died in February 1982 at the age of 45. After losing his job in Perth in 1964, he took a position with ABA, because it was the only employment he could find. The Armstrongs lived in the town with their three children and Ronald would often play with them after work in his overalls, which were always saturated with asbestos fibre. Julia Armstrong washed his work clothes, and she had no idea that she or the children or her husband were at any kind of risk from the fibre. She is now concerned about her children's health and the possibility that they may have contracted mesothelioma all those years ago. This is a fear among families who lived at Wittenoom; it is a constant gnawing anxiety.

While working for ABA, Ronald Armstrong developed a persistent cough and consulted a doctor several times, but there was no evidence of disease. He left Wittenoom in 1966 when the mine closed, and over the next ten years he experienced chronic ill health. He had contracted asbestosis, and he underwent surgery for cancer of the bowel, a disease associated with the ingestion of asbestos fibre. By the early 1970s, Ronald Armstrong was acutely ill and his wife was told by his physician, in an act of misguided kindness, that he would most likely die in his sleep. For those last months Julia Armstrong believed each time she said goodnight to her husband that she was saying goodbye to him forever.

Ronald Armstrong's illness badly affected his wife and their daughter who left school because she could no longer concentrate on her studies. Julia Armstrong suffered from emotional asthma, and during the final months of Ronald's illness she was under treatment for stress. The Armstrongs barely survived more than a decade of the anxiety of seeing a family member die slowly. The disease which killed Ronald Armstrong appears again and again in the life histories of the men who worked for ABA. There is nothing distinctive in the career of Ronald Armstrong, as a patient whose own health and family life were left in ruins by the stresses of enforced unemployment and of living with an illness for which there is no effective treatment. This reality is always absent from the medical literature describing the incidence and effects of asbestosis and mesothelioma. Neither is this reality part of the legal process which decides the rights of victims and their dependents for worker's compensation or damages.

Asbestos cement products were first manufactured in Perth in 1921 from imported South African fibre. Despite the pioneering work of Dr McNulty with mesothelioma, there was little concern and few publications from Western Australian physicians on asbestos-related disease until after the closure of the Wittenoom mine. In 1967 Dr Janet Elder, who appeared as a key witness at the Joosten trial, published a study on asbestosis, which was based on thirty-one cases she had reviewed over the previous five years.[13] All of Elder's subjects were male, and over half displayed finger clubbing, which is characteristic of severe asbestosis. Almost one-third of Elder's cases had died since first interviewed. The most striking feature of the subjects is the predominance of former Wittenoom employees: of Elder's list of thirty-one patients, twenty-eight had worked for ABA, whereas the other three had been employed in the asbestos cement industry in Perth. The former miners had periods of employment at Wittenoom, ranging from twenty months to twelve years, but the average period of exposure for those with no previous contact with asbestos was a mere seven years.[14] This figure is most disturbing as it is rare for asbestosis to appear in workers with less than ten years' employment. Elder comments that the premature development of the disease was probably due to a lack of ventilation in the mine and a lack of respirators.[15] The most striking of the cases cited by Elder is that of a man with severe asbestosis who had just over one-and-a-half year's exposure. At the very least, Elder's work suggested that the number of cases of asbestosis from the Wittenoom mine was certain to increase over the coming years.

As the number of cases of asbestosis and mesothelioma from Wittenoom continued to rise, so did the interest of Perth physicians in studying the diseases. In 1968 Dr James McNulty presented a paper on the subject of asbestos exposure at an international conference in Sydney.[16] The paper is unusual in that McNulty covered the subjects of both occupational health in industrial hygiene and the growth of medical knowledge. Most of the literature on the Wittenoom mine is confined to one or other of these areas, which are treated in isolation from each other. McNulty's paper is very different in that he gives a complete picture of what happened at the mine and why the levels of disease would certainly increase. By 1968, cases of asbestosis and mesothelioma had been recorded in every Australian state. The first asbestosis patient McNulty had identified from Wittenoom was a 33-year-old miner who also had tuberculosis. After surgery for TB it was discovered that he had asbestosis. This took place in 1958. In the period from June of that year until 1969, 103 men who worked for ABA had been diagnosed but, of these, approximately one-third had some exposure to dust prior to their employment at Wittenoom. Of the remaining group 41 were mill workers and 29 had been employed underground.

Given the conditions in the mine and the mill, it is surprising that no case of asbestosis was identified until McNulty's diagno-

sis in 1958. McNulty explained this delay as being due to the
tempo at which the plant was operated. Conditions had always
been poor and dusty, but it was not until production levels were
increased that the mine began to produce asbestosis in workers
after only a few years of exposure. In the McNulty sample, the
disease became apparent on average after 5.25 years among mill-
hands, and 6.6 years was the average duration for underground
workers. In the mine the fibre seams were six to nine inches wide
and the stoop was only a little over three feet. All miners were
exposed, not only to fibre, which in such a confined space was
oppressive, but also to silica dust which came from the drilling
of the host rock. Consequently, miners fared little better than
did mill-hands in terms of duration of exposure and the incidence of disease. Half of the men examined by McNulty displayed
finger-clubbing, and he and other medical officers were continually being surprised by the rapidity in the onset and progression
of disease. In some cases a miner's X-ray would be normal one
year, and at the next annual check he would display finger clubbing and basal crepitations. The work of physicians was made
even more difficult by the variable quality of the X-rays which
were done by a mobile van. In 1968 McNulty believed that the
future of many of the former miners was bleak, and he expected
the number of cases of asbestosis and cancer to increase in the
coming years. He also believed the outcome would be somewhat
worse for the mill-hands than for the miners, as the levels of exposure were highest above ground. McNulty's gloomy predictions
eventually proved correct.

As more and more cases of asbestosis and mesothelioma began
to emerge, the Western Australian public became aware of the
threat posed by exposure to any form of asbestos. This awareness was accelerated by growing publicity in the United States
and the United Kingdom about the industry's past and present
victims. In 1978 the Western Australian Public Health Department undertook the first major study of health at Wittenoom.[17]
The document presents a history of the department's involvement
at Wittenoom in the period from 1948. In essence, it was designed to absolve the department of responsibility for disease among
ex-miners, which was becoming uncomfortably obvious. The

document is useful, however, in substantiating claims about conditions at Wittenoom and in establishing the level of concern within the department at various times about the threat to the health of ABA employees. By 1959, the department was concerned about the mine and in particular about the number of young men who were developing asbestosis after only a brief period of employment.[18] The department cautioned ABA and local shire authorities repeatedly against the use of tailings, and a number of attempts were made to have the practice banned. At the time of the publication of Wagner's South African study on mesothelioma, the department's officers were already investigating asbestosis and silicosis at Wittenoom. In a memorandum in October 1961 from a senior departmental officer to Mr A. O. Allan, the then manager of ABA, the Department of Public Health set out clearly its fears about the mine. The letter states in brief: "This department is becoming increasingly concerned at the number of employees and ex-employees of Wittenoom who are presenting as patients suffering from asbestosis or silicosis. Mr McNulty informs me that he considers a number of dust counts at Wittenoom to be too high".[19] Following receipt of this letter a meeting was held in Perth between representatives from the departments of Health and Mines and officials from ABA. The company readily agreed to improve ventilation at the mine and to introduce a system of dust counts. The Department of Health also received assurances from the company that the use of tailings around domestic dwellings would cease immediately.[20] But it was not until 1966, the final year of the mine's operation, that the department was satisfied that accurate dust counts were being taken. By the time those counts were processed, the closure of the mine had been announced, and ABA ceased to be of any concern.

By 1977 the Wittenoom mine and its former workforce had become something of a *cause célébre* in Western Australia, and the prospect of researching a unique industrial tragedy attracted the attention of numerous physicians who realized the importance of what was happening among the ex-miners and their families. In 1980 the initial results from a five-year study by physicians from the University of Western Australia, financed in

part by a grant from CSR, were published.[21] The work was a
follow-up study directed towards the 6,200 former male employees of ABA. Within this group all cases of mesothelioma and
asbestosis diagnosed prior to 1 July 1979 and all deaths occurring before 1 January 1978 were collated. The majority of the
cases of pneumoconiosis were identified from the files of the
Pneumoconiosis Board and some others were gleaned from public hospital records. The researchers constructed an index of
exposure, distinguishing between those workers who had high,
medium-to-light and non-specified exposure in the mine. In each
of these three groups, the incidence of pneumoconiosis was
found to rise with duration of employment, with the highest
rates occurring in the most heavily exposed group. The incidence
of pneumoconiosis was found to be 4.8 per cent for these men. It
is important to remember, however, that the method used in
identifying cases of pneumoconiosis was imperfect and most certainly underestimated the actual numbers involved. Many
miners were migrants who reeturned to their country of origin
and the methods of diagnosis of the two principle diseases is also
likely to depress the actual figures. The figure of 4.8 per cent is
therefore nothing more than a conservative guess.

The researchers discovered twenty-six cases of mesothelioma,
exhibiting an interval period of exposure from thirteen to thirty
years. Not only was the latency period relatively brief, but quite
unexpectedly there were no cases of peritoneal mesothelioma.
As with pneumoconiosis, mesothelioma appeared to be dose-
related. In comparing the incidence of such disease there was a
significant surplus. The anticipated number of cases of lung
cancer in Western Australia for a group the size of the ex-miners'
cohort was thirty-eight, whereas the actual number of cases
identified was sixty. This suggested a twofold increase in lung
cancer among the Wittenoom group.[22]

A second major study of mesothelioma in Western Australia
was carried out by a group of physicians led by Dr B. Armstrong
and was released in 1983.[23] This study dealt with ninety definite
or probable cases of mesothelioma which had been diagnosed in
the period 1957–1980. It revealed an increase in the disease since
1974 and the researchers found a clear association with asbestos

exposure for almost all of the cases, with nearly half having been at one time or another resident at Wittenoom. Typically, the average time for survival once diagnosis had been made was less than a year, and in all patients the disease proved fatal. Like the earlier follow-up study, there was found to be a predominance of pleural mesothelioma, with almost 90 per cent of subjects being male. The period from the date of exposure until diagnosis ranged from eleven to forty-seven years, and in only eleven cases was there no definite association with asbestos. The researchers admit a problem in proving a causal relationship between exposure to asbestos and mesothelioma, owing mainly to a difficulty in defining what constitutes exposure. They concede that the nature of the causal relationships is little understood and that the issue of a dose–effect relationship remains unclear. As with the previous study the researchers had to rely upon hospital records as the principal means for identifying subjects, which were certain to have led to an underestimate of the actual numbers involved.

The problems of identification are cited in a further project carried out by a team from the University of Western Australia which published preliminary results in July 1984.[24] This study deals with the incidence of mesothelioma in the period 1960–1982 and, unlike the previous two works, a variety of means was employed in locating subjects. The authors admit that, during the first ten years in which mesothelioma was on the increase in Western Australia, it was most certainly underdiagnosed because of a lack of expertise and experience among physicians, who were unaware of the difference between mesothelioma and other cancers of the lung. Even so, in the period under review, 138 cases were identified. Of these 70 per cent had a probable exposure to asbestos. Fifty-six of the patients had been employed by ABA and another four had non-occupational exposure at Wittenoom. Sixteen had been exposed to blue asbestos at sites elsewhere in the state but, as with all previous studies, the authors found an under-representation of cases involving the peritoneum. The authors conclude that as a result of the operation of the Wittenoom mine there is an epidemic of pleural mesothelioma in Western Australia.[25]

The emerging story of asbestos-related disease in Western Australia has two distinct aspects. The first is the slow but steady growth in the medical literature from 1967 onwards which explores the consequences of the way in which the Wittenoom mine was operated. That literature is modest in that the authors invariably set out to achieve a small but specific goal, such as describing the numbers of cases and classifying them according to their various characteristics. The literature is not public in the sense that its audience is largely confined to specialists, and it rarely reaches a wider readership. The second aspect of the literature is found in the official figures on asbestos-related disease presented in parliamentary papers, often in response to questions in the House and in the annual reports of the state Department of Public Health. These official figures are in a sense more important than those in the orthodox medical literature, because they were intended for public scrutiny and provide a means for judging the level of public action over the issue. By 1980 there had been 195 cases of asbestosis and 64 cases of mesothelioma identified as resulting from Wittenoom, but the emergence of these cases stretched back over some twenty years.[26]

The first case of asbestosis in Western Australia was diagnosed in 1958 but, by 1961, the Commission of Public Health's annual report noted that another four cases had already been identified.[27] In the following year a further two new cases had been recorded and the commission responded by establishing a new section within its Occupational Health Branch dealing specifically with pneumoconiosis. The commission was aware of the connection between the disease and the dust levels at Wittenoom, and the annual report for 1962 contains the following comment: "It is planned to repeat the counts if necessary and advise management on measures of prevention of dust."[28] Despite this statement of intent, by 1963 the commission had come to view the incidence of asbestosis and silicosis as disquieting, and Dr McNulty had been appointed to take charge of the section of the Occupational Health Branch dealing with miners. The commission was also concerned about the use of asbestos in insulation and in particular with the practice of spraying the material into confined spaces. In the commission's own words, "A new method of spraying asbestos on ceilings during building construction exposes workers to heavy concentrations of asbestos dust and adequate respiratory protection is essential."[29] It is not known if the commission directly approached private industry about this problem but, in the light of the commission's behaviour towards ABA, it seems improbable that any action was taken in the building industry.

The annual reports for 1964, 1965 and 1966 reiterate the concern about the Wittenoom mine and the already-known consequences of high dust levels, but these protests have a hollow ring. By December 1966, ninety cases of pneumoconiosis from the mine had been identified and, of these, twelve had died. According to the commission it had been unable to persuade ABA of the importance of preventive measures to protect the men: "Although repeated warnings have been given by the Public Health Department over the years regarding the danger of asbestos dust, the industry has been slow to appreciate the situation."[30] Even though the mine was closed in that same year, the number of cases of pneumoconiosis continued to rise. In 1965 the Pneumoconiosis Medical Board awarded compensation for asbestosis to eight miners; in 1966 the number had increased to twenty and in 1967 it had risen to thirty-one.

From 1973 to 1980 there was a steady and relentless rise in the numbers of ex-miners from Wittenoom suffering from asbestosis and mesothelioma. By February 1978 there were 28 proven cases of mesothelioma, 89 cases of cancer of the lung and 198 cases of asbestosis.[31] By that date the commission was aware that the incidence of lung cancer among Wittenoom miners was 50 per cent greater than for comparable groups in the Western Australian population. The commission was also willing to admit that the number of such cases would increase still further as a result of the Wittenoom experience. This prediction was proven correct and, by 1982, asbestos-related disease had come to dominate the new cases of lung disease accepted by the Pneumoconiosis Board. In that same year there were 10 new cases of mesothelioma.[32] This increase must, however, be seen in the context of an Australia-wide rise in the incidence of this disease. By July 1984 the Commonwealth Institute of Health had recorded 550 cases of mesothelioma on its register and nearly 120 cases had been accepted by the Dust Diseases Board of New South Wales in the period from 1968. According to Professor Brian Gandevia, only a small part of this increase can rightfully be attributed to a heightened awareness by physicians.[33] In his view there had in the past decade been a genuine, dramatic rise in mesothelioma among all sections of the Australian population. In Western Australia, the reason for this rise is clear: it follows closely the rise in production levels at the Wittenoom mine some thirty or so years earlier. From 1943 to 1950, output at Wittenoom rose from a few hundred tons to more than 1,000 tons per annum. At the same time the size of the workforce expanded from 50 to 200 men. During the early and mid 1950s output continued to rise from 4,000 tons in 1954 to 7,000 tons in 1958. Once again the labour force also increased in size. Although the mine was always dusty, there was no significant disease so long as the levels of output remained low. It was the combination of poor ventilation and higher levels of production which brought about the epidemic of lung disease and mesothelioma among miners.[34]

The sudden appearance of mesothelioma and asbestosis in Western Australia follows hauntingly the rise in production from the mine. This relationships was for some time disguised because of the length of the latency period which, in the case of mesothelioma, may be as long as forty years. By 1970 the latency period had begun to expire and it was from that point onwards that mesothelioma and Wittenoom became a public issue. The reaction by the state government was predictably confused and haphazard, as there was no precedent for the range of problems which the mine's past and the town's future presented to a variety of government departments.

Since the issue of Wittenoom and the dangers faced by the remaining residents became a public controversy in 1977, the behaviour of state authorities has been quite erratic. The residents have sought to protect their property interests in the town, which would be ruined if they were forced to leave without adequate compensation, and the state government has been fearful of the possible long-term consequences if a health danger in the present town were allowed to go unchecked. The threat arises from the tailings which were spread about the town during the twenty years of the mine's operation. The tailings pose no danger of asbestosis, but there is the possibility that residents could contract mesothelioma. According to the World Health Organization, there is no level of exposure to blue asbestos which is safe, and that is a reality at Wittenoom of which governments have been aware. Unfortunately, there has been little awareness of environmental health issues in general within the state parliament.[35]

In November 1978 there was a strong reaction by Wittenoom residents over announcements that the town would be closed down. Under a cabinet decision all government employees at Wittenoom, including police and hospital workers, were to be withdrawn. Water and electricity supplies were also to be terminated, leaving Wittenoom a ghost town. The minister for Health, Mr Young, warned Wittenoom residents that the government was not legally obliged to provide any compensation for the loss of property or the devaluation of their homes and land. The residents faced the prospect of losing not only their community but also most of their tangible assets. It was clear from the minister's statement that within twelve months the town would cease to exist. The reason for this decision was the health hazard which specialists from the departments of Public Health and Mines agreed existed at Wittenoom. In a Department of Public Health study published in 1978[36] there is reference to the widespread use of tailings, which was still evident years after the mine's closure. The document contains the following observation: "The race course is completely covered with tailings and it is interesting how the fibres have become pulverized and separated with rain to form concentrations in the hollows made by footmarks."[37] In July 1977 airborne-fibre samples were taken by Dr Cumpston and Mr D. Sykes. The ten samples collected from a car at various locations about the township had an average reading of a little under 2 fibre/cc. The presence of fibres was also detected at the primary school and in the dust taken from vacuum cleaners in domestic dwellings. Dr Cumpston was certain as to the presence of airborne fibre, but he was less convinced that the fibre presented a demonstrable risk to health. He recommended, however, that all future growth in the town be discouraged and that samples be taken annually. This was done again in July 1978, during which Dr Cumpston explained to the community at a public meeting the possible risk they faced by choosing to remain at Wittenoom. In particular, he pointed out the risk to the children whose life expectancy would be sufficient to absorb the latency period involved with mesothelioma.

In November 1978 at a public meeting attended by most of the town's residents, the community voted to withhold payment of rates until they were given a promise of government assistance. They claimed that they had been guaranteed help more than six months earlier and that the promise had not been honoured. They also demanded that the existing tailings be covered over as a solution to the problem of environmental contamination. Lang Hancock, the major property holder at Wittenoom, was confident there was no risk to health. In the week of the release of the Public Health department report, *The Health Hazard at Wittenoom,* Hancock explained: "on balance, Wittenoom, with the exercise of a little common sense, can be made a safer place for children than any capital city."[38] Hancock also suggested that with the correct approach Wittenoom could become a major tourist centre and that one day the town could boast an international airport serving the Pilbara region as a whole. Hancock's optimism did not deter the Public Health department from completing a further study, *Exposure to Crocidolite at Wittenoom.*[39] This survey was based upon the use of long-term personal samplers which were worn by residents over a six-hour period. In this way seventy valid samples were obtained. Several of the readings were in excess of those levels permissible in an industrial setting.[40] The authors of the report point out that, unlike a work environment where the worker is for most of any day absent from the source of contamination, for residents exposure is constant. The report concludes with the finding that there is a perceptible risk to health, especally for the children, which cannot be ignored.[41] In a statement of 3 April 1979 addressed to the people of Wittenoom Mr Young, the minister for Health, referred to the study as proving beyond doubt the hazard, especially to the town's children. At that time it appeared that the fate of the township had been decided, because there could no longer be any debate about the risk to health. That, however, was not to be.

In a joint press release by the Western Australian government and the Wittenoom Working Committee in August 1979, it was announced that the town would not be phased out. After meeting with the committee, which represented the township, the minister for Health accepted that there were many residents who were committed to staying. As a result of the meeting, it was decided that three individual committees would be established to monitor health and welfare, and also to investigate ways in which the environment could be made safer. One committee was also to be devoted to the development of tourism. The minister made it clear, however, that any person wishing to leave before 30 June 1980, would be given government assistance, and that no government employee would be forced to live at Wittenoom. The sentiment of the news releases is conciliatory, with the government bowing before the intransigent stance of the community and the impossibility of either covering the existing tailings or of persuading the people to leave voluntarily. The final paragraph of the release contains the comment: "It was agreed that the future of the town was of paramount importance, and that what happened in the past should not be dragged up."[42] Presumably this refers to the known danger from airborne fibre described in Public Health department reports, and to the previous decision to close down the township. There is no justificaion given in the statement for the government's about-face. However, the issue was soon raised to the state parliament.

In a statement before the House on 10 October 1979, Mr Davies, the then leader of the Opposition, referred to the danger at Wittenoom and to the need for the residents to be given adequate warning about what staying in the town would mean to themselves and to their children.[43] He also commented that the actual degree of hazard was not understood, and that there was an urgent need for further research. Mr Davies pointed out the lack of proper documentation about state-owned properties at Wittenoom, which was disturbing, given the controversy which had surrounded the town for some years. He had approached the Education department, but it had been unable to give any indication as to the value of school property in the town. He was able, however, to discover that the hospital was valued at over $1 million. The state Housing Commission still owned sixteen houses valued at $129,000, and the Department of Public Works was unable to say whether or not it owned any properties. At the least, Mr Davies' investigations had revealed a lack of knowledge among the various government departments which shared some immediate interest in the fate of Wittenoom.

Mr Davies once again raised the issue of Wittenoom in the House on 6 December, because of growing public concern and because of pressure from Wittenoom residents. He demanded that the government not withdraw essential services until a comprehensive study of the fibre levels had been completed.[44] He also suggested that residents should be given the option of staying if they chose and, to this end, the government should investigate the tourist potential of what he described as a "beautiful town". The state government however, continued with its policy of encouraging residents to leave voluntarily, and some assistance was given to those who chose that path. It also maintained its decision to refuse to expand existing essential services, so that no new homes could be built and the population would continue to dwindle.

By August 1981 there were only 161 residents remaining at Wittenoom. Government employees continued to be posted to the town, but only those without children and only if they agreed to the transfer. The policy adopted by the regional administration for the northwest of refusing water and electricity supplies to new residents has been effective. There is little employment, apart from three small prospecting firms, one of which is owned by Hancock and Wright. The Wittenoom residents did, however, have some success in pressing for the development of a tourist industry when, in August 1981, Mr Lawrence, the minister for the North West, announced plans for a privately financed tourist development resort.[45] The new development was to be located some six kilometres from the existing township, and essential services were to be extended from there to the new centre. Mr Lawrence explained that it was not feasible to render Wittenoom safe by covering the existing tailings, although the government had already provided some funds. The new tourist development was intended to replace the town and there were, according to the minister, no immediate plans to compensate residents for the loss of property if they decided to move. The decision was greeted with a mixed response. Some residents objected to what they saw as the destruction of their community, although the tourist project did suggest the possibility that the town would survive in the long term, if only as an auxiliary centre. The demands by residents for the sealing or removal of the tailings had always been viewed by the Western Australian government as unrealistic. In March 1982, a Public Works department report estimated the cost of such work to be in excess of $3 million, a sum which could hardly be justified for so small and isolated a community in a time of economic restraint.[46] There are also an estimated two million tons of tailings located on private land near the town. These tailings are not fenced and they are not monitored by state authorities which view them as part of a private dump.[47] It is possible that materials from this dump are being washed by rain storms into the Joffre Creek which, in turn, feeds into the Fortescue River. That river runs into the Mill Stream, from which part of the water supply for the Pilbara region is drawn.[48] The minister for Health is aware of this problem, but he is convinced there is no threat to the health of the people consuming asbestos-impregnated water. He has never explained how he arrived at such a conclusion.

The erratic history of what is to be done with Wittenoom continued throughout 1982 and 1983. In July 1983 the Western Australian government announced an $85 million plan for the development of a new tourist complex.[49] By that date only eighty-five people still remained at Wittenoom, including seventeen school-age children. Under this new plan the residents were to be given priority for employment and in establishing new businesses at the site, which will be located in the Hammersley National Park, some forty kilometres from the existing township. Once again the reaction from residents was mixed, because of the way in which the issue has been handled. The emerging medical literature and studies by the Department of Public Health did little to allay fears that life at Wittenoom could ever be safe. A study by Drs Musk, Baker and Whitaker published in 1983 found no evidence of illness among residents, but the researchers did not exclude the presence of a health hazard.[50] Such research has done little, however, to persuade residents that they would be better off living elsewhere. For the state government, Wittenoom has continued to be an embarrassing reminder of an unpleasant past which refuses to go away. The township, like the dying miners, is a reminder in a state in which development is a ruling ideology of the costs which follow the failure of government to regulate and control the activities of private industry.

The interests and needs of the former miners have been represented by the Asbestos Diseases Society which, over the past four years, has been so successful in keeping the plight of these men and their families before a largely disinterested public. The society was founded in November 1978 by a senior Canberra public servant, Trevor Francis, who, for a brief period in the 1950s, had worked for an asbestos company. Mr Francis now suffers from asbestosis, the disease from which both his parents died. In launching the ADS, Mr Francis sought to provide a means of help for the victims of asbestos disease and also to present a counter to the industry's constant and effective propaganda. From its inception the ADS has been funded by private donations, and it is committed to expressing the viewpoint and interests of the concerned layperson. The society has been represented at various public inquiries, such as the Williams Inquiry in New South Wales and the federal inquiry into hazardous chemicals.[51] In these forums the ADS has given the victims a voice and, thereby, has forced industry to respond to serious questions about its behaviour and responsibilities. The ADS has had some success in petitioning members of the federal parliament, and many questions asked in both Houses during 1979 and 1980 arose from the society's work. The ADS has never relented from pointing out the lack of social responsibility shown by all sections of the asbestos industry in this country. It has also highlighted the ways in which the industry's victims have been treated with insensitivity and neglect by government. In particular, the ADS has highlighted the legal obstacles, which victims face in seeking redress and compensation. Most important of all in the political arena, Trevor Francis and the ADS have been successful in helping to place the asbestos issue where it rightfully belongs: that is, out of the hands of technocrats and specialists and into the political arena.

In Western Australia the ADS has played a more direct role in helping the victims of asbestos disease and, in particular, the former employees of ABA. In 1979 the campaign against the asbestos industry gained momentum in Perth with visits by the journalists Timothy Hall and Matthew Peacock. Both attended the inaugural meeting which established the ADS in Perth after contacts had been made with Trevor Francis in Canberra. The creation of a local branch had taken some time to achieve, but the ranks of the victims from Wittenoom and from Perth asbestos cement plants began to swell. The attendance at the first meeting was a mere twenty people, but the numbers increased as more and more sufferers became aware of the origins of their illness and of the society's existence. By 1984 the ADS's regular meetings were being attended by fifty to sixty people and the annual general meetings usually attract over a hundred members. The major function of the ADS is to counsel victims and their families about their rights to compensation and medical care. Such work is vital among people who are poorly educated and unable to articulate their rights. Many of the people who come to the ADS offices for advice have already reached the point of selling off their chattels in order to survive. This situation improved somewhat in 1983 with the introduction into parliament of the Asbestos Related Disease Bill but, for so many, the loss of the earnings of the family breadwinner has brought poverty.

The organization is now known as the Asbestos Diseases Society of Australia Inc. The ADS defines its role as being to combat not only the companies which produced asbestos but also the trade unions and government, which it sees as having done nothing to help the victims. It is, therefore, a unique organization in boasting no ready allies, and enemies in three powerful camps. Many of the clients who come to the ADS are people who have been unable to receive any form of assistance. Often men with asbestosis and mesothelioma have failed to gain a clear explanation from physicians about what the disease means and what alternatives they may have for the future. When clients arrive at the ADS offices they are given an initial interview during which a thorough work history is recorded. The centre will arrange medical appointments, if necessary, with the Respiratory Section at the Queen Elizabeth II Medical Centre, where specialists in asbestos-related disease are to be found. The diagnosis given at the Centre is reliable and provides better medical care than is available from private practitioners. This Centre will, if necessary, refer patients directly to the Pneumoconiosis Medical Board for assessment. If the patient's case is not accepted by the Board the ADS will then arrange for legal representation, although patients are often loath to take such action. Many of the people who approach the ADS are reluctant to seek medical help and only do so when they are acutely ill. Like so many victims of industrial malpractice, these men and women tend to blame themselves for their illness. Where a case reaches the Workers Compensation Board there are preliminary hearings. These are followed by pre-trial negotiations and, if these are successful, a settlement will be achieved. But where a case is contested, which is quite common, the case can become drawn out. To the end of 1984 the most successful case achieved a settlement of $73,000.

If viewed in terms of lost wages over a ten-year period, this was not a generous result. When a pension award is made under the Workers Compensation Act, the financial problems faced by the victims usually continue. If married, a man can early only $60 per week before deductions are made from his pension. In the case of a single man, the figure is set at a mere $30. In both instances the result is chronically ill men supporting families on an income below the poverty line. For most sufferers that is merely a continuation of years of financial struggle. The declining capacity of the breadwinner to work has usually already eroded all savings, and the family may have been living in poverty before any compensation has been awarded. The burden of illness is but one of the myriad problems faced by asbestos-related disease sufferers. Most men and their families endure living in debt under the combined stress from ever-worsening health and ever-deepening financial hardship.

In Canberra, where the society was founded, the ADS has played essentially a political role in promoting the need for public awareness and effective scrutiny of the asbestos industry. This it has done through a variety of means, including the tending of submissions to numerous public and parliamentary inquiries.

Most of the submissions made by the ADS have been prepared by Trevor Francis and they are unusual documents. In these submissions, Francis has succeeded in asking the industry and its representatives a series of important questions. Those questions concern the gap between the industry's rhetoric and its actual practice in the work and market places. The gap is rarely, if ever, alluded to in government reports, and at the federal and state levels the Departments of Health have, as yet, been unable or unwilling to make manufacturers, such as James Hardie and CSR, accountable for their past actions. The absence of these questions has allowed the industry to purvey the myth that asbestos should be discussed only by experts, and that neither the victims nor the general public has the right or the need to know the details of the controversy. The ADS has challenged that view.

In a submission before the 1982 federal parliamentary inquiry into hazardous chemicals the ADS set out the political context in which the asbestos issue is embedded.[52] Trevor Francis refers to the arguments tendered in its own defence by the industry and to the *conceptual lens* through which those arguments are presented. Quite correctly, Francis claimed that the industry has always denied the existence of such a *lens*. The industry contends that its position is purely objective and is based upon scientific evidence. It is only among the industry's critics that arguments are infected with political aims and material purposes. That, at least, is the view of companies such as James Hardie and Johns-Manville. In truth, the costs involved in the regulation of industry have been trivial when contrasted with the eventual costs in terms of ruined lives, environmental pollution and the burden of public health care for the victims. As Mr Francis points out, both government and industry are anxious to ignore such a means of accounting. Industry is also able to ignore the multiplicity of exposures to fibre for workers and consumers. Very frequently the presence of asbestos in a product is disguised, so that consumers are brought unwittingly into contact with fibre. Much asbestos has been used by small businesses, which invariably are the least well-regulated part of any industry, and the numbers of heavily exposed workers must continue to mount. In the past, the risk associated with asbestos was acceptable, only because of the passivity of governments and the ignorance of the general public which is unaware of the dangers, known all too well to producers. Industry was allowed to decide what constituted the public good and what level of risk to employees and consumers was acceptable. Industry, of course, viewed the acceptability of risk as being part of a trade-off between its own special economic interests, which it equated with the common good, and the potential loss of profits which would result from morally responsible behaviour. This meant that the costs for the damage done to the health of workers, to the welfare of their families, and to those people, including children, who came into contact with asbestos have been shifted away from the companies and on to the shoulders of the community at large. The ADS is correct in its claim that the issue of asbestos and health is, above all else, a political issue which must finally be arbitrated by governments and their public, and not by the companies.

6

Baryulgil: The Mine

Baryulgil is a small and isolated community in the Northern Rivers region of north-central New South Wales. The township is some eighty kilometres by road from Grafton. Ironically, Baryulgil and its people have enjoyed a more gentle history than that experienced by other Aboriginal communities in that state.

The first contact between Aborigines and Europeans in the region occurred during the 1830s with the influx of sawmillers in search of the rich cedar forests. The sawmillers were soon followed by the pastoralists. During the 1840s local Aborigines reacted violently to white settlement and there followed successive periods of armed resistance and retreat into the mountains. In the periods of quiet, the pastoralists moved further up the Clarence, and the rich lands adjoining the river were quickly settled.

Despite the relative shelter afforded by the isolation of the district and the late arrival of European settlement, there were a number of massacres of Aborigines in the Clarence River area. In April 1841 a shepherd was speared by blacks near the Yulgilbar station. In the company of the notorious Border Police the station's owner, Edward Ogilvie, led a reprisal raid. There were also other incidents in which children, men and women were killed at Purgatory station and Tabulum in 1841, and a massacre of blacks on the banks of the Clarence took place as late as 1852. According to Prentiss,[1] during the entire period of violent contact the numbers killed among the Bandjalang and Kumbainggiri tribes could have been as low as one hundred. Unfortunately, Prentiss gives no indication as to how he arrives at this figure.

There were, however, fewer than twenty deaths of Europeans at the hands of Aborigines. This early period of armed resistance soon passed, and after 1846 killings of Europeans were rare. For the black communities of the Clarence a far greater threat to survival came from disease and from the loss of traditional land.

From 1849 onwards, Aborigines were employed as casual labourers by the cedar men, and in the boiling down of sheep and cattle, in harvesting corn, cutting bark and in work as stockmen. Opportunities for such employment improved during the next decade as many whites left the district for the goldfields. Increasingly, responsible jobs were taken by blacks who now worked at mustering, as shepherds and in the building of stockyards. A few Aborigines also joined the ranks of the Native Police.

The Bandjalang people of Baryulgil first came into contact with Europeans when Edward Ogilvie arrived to establish a pastoral empire along the banks of the Clarence River. Ogilvie befriended the local tribes and he readily employed blacks on his station, which he named Yulgilbar. Ogilvie learned the Bandjalang language and is credited with producing its first grammar.[2] He is also credited with having behaved in a humane, if paternalistic, manner towards the local communities which were relatively safe until the influx of miners in the brief gold rush of 1871 when gold was discovered at Solfernio. The Land Census of 1841 shows Ogilvie as the largest landholder in the Clarence River district with Yulgilbar having a river frontage of over fifty-six miles. Ogilvie set about building himself a castle at Yulgilbar as a symbol of the permanence of his pastoral empire. For this he imported stonemasons from Hamburg who constructed a home containing forty rooms, with extensive quarters for servants and guests. At the centre of the building was an internal courtyard featuring a marble fountain made in Florence and this, with the twin towers set at either side, gave the house a Moorish appearance. The castle also boasted an Italian crystal chandelier and it is estimated that on completion Ogilvie's new home had cost £40,000 — a vast sum for a building constructed from local materials and using, for the most part, cheap labour.

Mrs Lucy Daley, a surviving member of the Baryulgil community, remembers as a small girl working in Ogilvie's castle where her mother was employed in the laundry.³ Wages paid at the castle were good and domestic service offered regular work. Those blacks who were employed as stockmen were paid only a small wage but they were also given generous rations. During the period from the arrival of Ogilvie to the end of the First World War, the Bandjalang people lived near the castle in huts by the river, on what had always been traditional land. Ogilvie set aside this area as a reserve. The people bought supplies such as tea, sugar and bread from a store at the present site of the Baryulgil Square, which is some ten kilometres from the original camp. Mrs Daley remembers that in 1918 it was suggested to the people that it would be best if the settlement were moved closer to the store and away from the river. The oldest members of the community attended a public meeting with the then manager, Mr Morrissey, who assured them that, if they moved, they would be guaranteed the right to the land for ever. The Bandjalang people had little choice other than to agree.

In moving to The Square the community left a permanent water supply, for the creek running adjacent to the new camp was unreliable. This made life more difficult especially for the women, as water for washing, cooking and bathing children had to be hauled from the often-dry creek, or else these tasks were carried out on its muddy banks. Proximity to the store was little compensation, but the move did give the station freer access to the river flats. It also allowed the Bandjalang the opportunity to remain unfettered by mission life and thereby to retain a degree of freedom unique for a black community in New South Wales during that period.

In 1918, the year of the move, asbestos was discovered at Baryulgil when a black named Bill Little led prospectors to an asbestos outcrop. For this he was paid a bottle of rum and a plug of tobacco. Men were wanted to work in this mine, but the demand for labour was never great and the mine lasted only a brief period, closing finally in 1924. It is not clear if the removal of the community to The Square was orchestrated in order to provide labour for the mine or if the two events were simply coincidental. The move, however, did nothing to change the economic dependence of the community on the Yulgilbar station which had no choice other than to follow the advice of the manager, Morrissey, and settle at The Square. To the benefit of the community, this relationship with the station meant that the people of Baryulgil were never subject to the control of a white reserve manager, and they were thereby able to retain a large degree of cultural independence. The people were also able to continue living on what was then, and remains today, traditional land. As in the past, at Christmas the community would move to the banks of the Clarence where as many as two hundred people would gather to fish for eels and turtles. With the decline of Yulgilbar, however, the demand for labour fell dramatically and so did the size of the Baryulgil community as young men left the district in search of work. By 1943 the official population for the reserve, as it was then listed briefly, was a mere fifty-three persons.[4]

The Baryulgil Mine

Chrysotile or white asbestos occurs in serpentine rock in a seam less than one kilometre southwest of Baryulgil post office. During the initial period of its exploitation, that is, from 1918 to 1924, only 2,500 tons of fibre was recovered. The company which operated the mine at that time was named Asbestos Mining Company of Australia and it failed principally in the face of competition from imported South African fibre. There is some evidence that James Hardie was directly involved in this initial venture and it is almost certain that the fibre itself found its way into Hardie products. In 1940 Wunderlich Ltd, which was later to become a subsidiary of CSR, began development of the same deposits. In 1941 the company extracted 144 tons of fibre which was used in the manufacture of asbestos cement sheeting. In the following year output had fallen marginally to 103 tons valued at just over £4,000,[5] but in 1943 production increased significantly and the Yulgilbar district thereby became the chief domestic source of asbestos fibre in the state. According to the Department of Mines,[6] open-cut mining at Baryulgil produced ore graded initially at 2–3 per cent which, with careful handpicking, could be improved two- or threefold. The quality of the fibre was suited ideally for use in asbestos cement products. In 1943 a small plant was installed, and with it came further improvements in output. In the following year James Hardie became directly involved in the venture. The Hardie company was anxious to secure a local source of fibre because of problems in obtaining a regular supply during wartime from South Africa and Canada. Hardie formed a partnership with Wunderlich and established Asbestos Mines Pty Ltd with each partner holding an equal share in the company.

Conditions at the mine during this early period were primitive. Initially the ore was sorted and bagged by hand and then trucked along the unsealed road to the railhead at Grafton. The sorting of ore was done by women who were employed on-site for the first year of the mine's operation.[7] There was no crusher at the mine and processing of the ore was done at a mill in Brisbane. In early 1942 a small crusher and separator were introduced. The fibre was separated from the ore body and collected in a fibre room. Ken Gordon, who worked at the mine during this period, remembers that at the end of each day a team of men would go into the room and shovel up the fibre which was then bagged by hand. The room soon filled with dust and the men would often have to go outside the bagging section to catch their breath. This practice lasted for the next four or five years. The mill itself was very cramped and all the crushing and separation of the ore was carried out within a single building.

From 1944 to 1953 Wunderlich retained control and responsibility for the day-to-day operation of the mine. In 1953 the James Hardie group purchased Wunderlich's 50 per cent share, and from then until 1976 Asbestos Mines Pty Ltd was operated as a wholly owned subsidiary of the Hardie organization.[8] From 1976 until its closure three years later the mine was owned and operated by Woodsreef Mines Ltd after it had purchased all shares in Asbestos Mining.

The mine site was leased to its various operators by the Yulgilbar station. The opening of the mine in 1941 is remembered at Baryulgil as a welcome event providing a chance for much-needed work. Over the next thirty-five years, virtually all of the workforce at the mine was drawn from local Aboriginal communities. The prospect of regular employment, which is always difficult for Aborigines to find, attracted miners from the reserves at Tabulum and Muli Muli. It also brought in men from Woodenbong, Casino, Yamba, Grafton and Kempsey. It was common for men from Tabulum and Muli Muli to stay at The Square during the week, returning home on Friday evenings. For Baryulgil, the mine allowed the community to break away from the chronic unemployment which was the shared fate of other Aboriginal communities in New South Wales. It also gave the people a feeling of pride and independence in being able to do work which was difficult and often highly skilled. The only Europeans employed at the mine were the manager and the fitter. All other jobs were performed by Aborigines. They worked with jackhammers, as mill-hands, as powder monkeys and as drivers. They did repair work on the machinery and they laid the benches in the quarry. It was Aboriginal labour which did the bulk of the work in the construction of the new mill that was completed in 1958.

Although the workforce at Baryulgil never numbered more than forty men, there was a time when the mine was promoted as a possible source for the economic development of the north coast region as a whole. In an article published in a local paper in 1961,[9] it was claimed that Baryulgil could become one of the greatest asbestos mines in the world. Supposedly, Baryulgil held the prospect of vast underground wealth in an area boasting copper and iron as well as asbestos. The article goes on to claim that there was enough fibre in the region to provide the stimulus for other mining ventures which could bring in their wake much needed investment and employment. The paper also features a photograph of the newly completed mill which is cited as costing more than £50,000.

The mine at Baryulgil was always worked as an open-cut quarry with benches cut down both sides of the pit. Drilling was done with jackhammers, and explosives were used to dislodge ore from the quarry face. Ore pieces the size of kitchen tables were broken up with sledge-hammers. The richer ore was placed in skips each holding around 1.5 tones. When full, these skips were emptied and the contents trucked up to the mill. The extraction of ore from the quarry was always the most labour-intensive part of the mine's operation, with between twenty-five and thirty men working in two teams. No work clothes were issued by the mine and boots were made available only during the final years of its operation. The work was hard, with shoeless men breaking ore with 14-pound sledge-hammers. A former miner, Neil Walker, recalls that each man was required to fill ten bins per day.[10] In summer the work was particularly demanding as the miners had also to sort the ore before it was trucked to the mill. This handsorting helped to upgrade the ore from a 2–3 per cent yield of fibre to as high as 8 per cent.

Following blasting, the quarry filled with a cloud of dust and fibre which drifted out over the nearby mess area.[11] The dust covered everything and it was usual for the quarry workers to re-enter the site before the dust had settled. In the Old Mill, which was in operation from 1944 to 1958, no attempt was made to suppress the dust, and at the end of each day's work the fibre was simply shovelled by hand into hessian bags. Former mine manager Jerry Burke remembers that the men engaged in bagging during this time at Baryulgil were Andrew Donnelly, Harry Mundine, Benjamin Oba, Richard Mundine, Albert Priest and Joe Waghorn. All these men are now dead.[12] In the mill there was no air filtering or effective dust-extraction system which meant that dust was a constant nuisance. Burke recalls, "The old mill was such that when you walked in it was impossible to see anywhere. Even the operator standing beside you was practically invisible."[13]

In 1958 a New Mill was built based upon what the company now chooses to refer to as "an advanced design". As with the old plant, milling and processing of ore involved three basic operations: the winning of the ore, the sorting of ore at the quarry face and the treatment of ore in the plant to extract the fibre from the rock. The ore was trucked from the quarry to the mill and emptied into a crusher which reduced its size. The broken ore was then conveyed through a series of milling machines and the asbestos fibre and dust was drawn off by an exhaust system to separate them from the gravel. The dust and fibre were fed into cyclones for final extraction. The recovered fibre was then bagged. The residue dust was fed into a dust room where it was collected in a series of suspended socks. There were forty to fifty of these flanellette socks lined up in rows from the ceiling of the dust house. During the later phase of the mine's life, calico socks were introduced. In either case the socks soon wore out, and when clogged would belch dust and fibre. One of the least pleasant jobs in the mill involved entering the dust house and tapping the socks to maintain a flow of material. There are numerous references in the company's inspection reports throughout the 1960s to the need for this process to be mechanized. The serpentine dust was bagged for use in paint manufacture.

The bagging of the fibre was carried out at the bottom of a cyclone. According to the company, the 50-pound bags used for this purpose were made of hessian. In the mid-1970s, plastic liners were inserted to reduce the escape of fibre from the loose weave.[14] In the early days, the bags were sewn by hand but this was subsequently done by stapling. At any one time there were only a small number of men working in the mill and there was little interchange between mill and quarry. One of the five permanent mill-hands had the job of cleaner, and he would, according to the company, use a vacuum in clearing dust from benches, floors and walls.[15] This claim, however, is not supported by evidence from any of the former miners whose testimony on working conditions in both the Old and New Mills, given before the House of Representatives Inquiry into Baryulgil held during 1983 and 1984, is in sharp contrast to that presented by James Hardie. The company, however, has consistently rejected such evidence as mere hearsay which is open to the kind of distortion characteristic of rumour and gossip.

Bill Hindle, a former fitter who worked at Baryulgil from 1954 until its closure in 1979, gave evidence before the inquiry.[16] Hindle began work in the Old Mill, which he remembered as being very dirty. According to Hindle the company agreed to a New Mill because it wanted not more fibre but a better product, free of gravel. But when the New Mill was established production levels were increased so that many of the problems found in the original plant were reproduced. There was also, he recalled, little attention paid to containing dusty areas or sealing off trouble spots within the plant, and the dust pipes used for extraction were inefficient. To clear the building of dust would have requir-ed constantly shutting down the plant which, of course, would have caused production to fall. Even so, the plant was stopped every three or four hours in an eleven-hour day to "clean down the dust bags".[17] Bill Hindle died from mesothelioma in Decem-ber 1984.

The machine which dried the ore as it came into the mill was always a problem. The dryer from the Old Mill was installed in the new plant where it was expected to handle almost double the capacity.[18] If the ore was not properly dried the plant would fill with steam which, when mixed with the airborne fibre and dust, would soon clog the exhaust ducts. The actual capacity demand-ed of the New Mill is uncertain. Hardie claims that the mill was designed to carry 6 tonnes of ore per hour and that at no time in the period from 1967 to 1976 did the level exceed 6.5 tonnes.[19] Many former miners recall that the mill was invariably overload-ed, thereby increasing the levels of dust. Evidence from Hindle and former manager Jerry Burke suggests that the design of the plant was inadequate. It is clear from interhouse correspondence from the mine to James Hardie's head office in Sydney that the company was aware that overloading of the mill increased the presence of dirt and stones in the final product.[20] This corres-pondence also indicates that the mill was designed to handle an output of 6 tonnes per hour.

Whatever the cause, it is certain there was a persistent problem with airborne fibre throughout the life of the mine. Jerry Burke claims to have spoken frequently to management about the dust but only after 1969, following the death of a miner, was any-thing done.[21] Burke was told that the mine had but a short life which was the reason the company was unwilling to spend the necessary $70,000–$80,000 necessary to install an effective dust extraction system. The few measures which were taken, such as the installation of a water spray at the tailings hopper, were not successful. The spray failed to function properly because of a lack of mains water pressure. In the bagging section the dust was oppressive. According to Burke, breathing in the section was like swallowing chilli.[22] Even so Burke is adamant that the new plant was a marked improvement on the Old Mill.

Testimony from men who worked in the mill is uniform and gives a picture of filthy conditions and a complete absence of concern with what are today accepted as normal hygiene practices. Miners and manager alike shared the same work environment. In evidence before the Baryulgil inquiry Jerry Burke recalled, "I was working as manager and I got covered in dust just the same as everyone else. Within an hour or so of starting in a new set of overalls you would be coated in fibre and dust, and the fibre even used to get in your socks and into your skin so that you would probably end up with half a dozen fibre splinters underneath your toes."[23]

Evidence from visitors to the mine confirms that Baryulgil was an unpleasant place in which to work. Rodney MacBeth, an organizer with the AWU, remembers that the town itself was shrouded in a perpetual cloud of dust. "The dust even on still days emanated from the treatment works and settled on everyone and everything in the vicinity."[24] In such an environment it was not only the miners but also their families who were at risk from the airborne fibre.

Greville Torrens, who worked at the mine for a number of years, recalls that the manager, Jerry Burke, was always followed about by a kelpie dog.[25] The dog spent twelve years at Baryulgil and it was frequently in the mill where the dust was most heavy. During the last years the animal became ill and terribly thin before finally wasting away. Burke claims he had an autopsy performed on the dog which revealed the adverse effects of asbestos fibre and the presence of asbestosis. The dog died in 1980, the year after the mine was finally closed.

Working conditions at the mine were unpleasant, which is not surprising given the smallness of the plant and the isolation of the community from trade union influence and from public scrutiny. Ken Gordon was 13 years old when he started work as a "billy boy".[26] This was the normal route for new miners from The Square and at 15 he went into the quarry to work on the skips, the job reserved for the youngest miner. Gordon would push skips containing waste and poor rock out of the quarry.

The quarry itself was always wet and pumps were kept running constantly to clear water from the bottom of the pit. The dampness often caused problems with the dryer as the ore had to be dried thoroughly before separation of the fibre. Gordon recalls that when he started work the miners did not wear hard hats and most tied a rag or handkerchief about their foreheads to stop the sweat from running into their eyes. It was the manager, Charlie Lloyd, who in 1969 first introduced safety equipment such as hard hats and rubber boots.[27] There were no showers available for the miners until the final years and the men used to wash after work in the creek. When showers were installed there were only four units and there was never sufficient hot water, which in any case was hard and difficult to bring to a lather. One former miner named Harrington recalls that, after each day's work, "Your skin was still white. You would wash it off and you would go like that afterwards and you would be a black fella walking along with a big white streak."[28] There was no canteen and the lunch room was an open-sided shed which was built in the mine's last years. Apart from the dust the work was made difficult by the heat and the constant noise from the machinery.

The mine was a small and insignificant part of the James Hardie conglomerate which explains some of the idiosyncratic efforts towards improving industrial hygiene. Charles Sheather, who worked at the mine from 1969 to 1972, recalls one attempt made to reduce dust levels in the bagging section, where the fibre was released from a cyclone and bagged.[29] The manager responded to complaints about the dust by installing an ordinary Expelair kitchen fan next to the cyclone. He purchased two such fans, the second of which he used in his own kitchen. Respirators were not generally used and the men working in this section were exposed to high levels of dust which was an inescapable part of their daily work.

Work in the quarry at Baryulgil was hard and held little appeal to local European labour. One white, an immigrant Scot, worked for a brief time in the quarry. After only a few days he collapsed from the heat while drilling on one of the top benches, even though it was relatively cool so far above the pit floor. Neil Walker remembers seeing the man swaying and about to fall.[30]

Walker and another miner grabbed the Scot, placed him in the back of a tiptruck and took him at once to the office. The manager's wife wrapped him in blankets and placed his feet in a bucket of ice. He was by this time frothing at the mouth and spent the next week in Grafton Hospital. The Scot did not return to work in the mine and apart from the manager and the fitter no European stayed for any time at Baryulgil.

It is difficult to understand how a major Australian company could have owned and managed the kind of plant that is portrayed in the oral evidence from former miners. Jerry Burke and the Aboriginal Legal Service, in their presentations before the inquiry, argued that demands for improvements at the mine were thwarted by a lack of decision at head office as to the mine's future. The chain of command governing the management of Asbestos Mining Pty Ltd is quite clear. The local manager would communicate directly with the technical director at head office in Sydney. On questions involving capital expenditure there would be a discussion of policy at Board level. Day-to-day decisions would be made by Mr Frank Page whose name appears on most of the interhouse correspondence between Sydney and the mine. Mr Page was a member of the Boards of Asbestos Mining Pty Ltd and James Hardie. Frank Page resided in Sydney but he paid frequent visits to Baryulgil and he was in regular correspondence with the local manager.[31] From 1953 Baryulgil was operated as a fully owned subsidiary of the parent organization, and whatever took place at the mine must have been known at head office.

Prior to 1953, James Hardie had little control in the running of the mine, which was viewed by the company solely as an investment to secure a supply of fibre in case overseas sources should be interrupted. After Hardie purchased Wunderlich's interest in the operation this relationship changed. With the change in ownership in 1953, the existing manager, Peter Neilson, was retained. He was followed in turn by Allan, Lloyd and Burke, all of whom worked in close contact with head office in Sydney. Unlike smaller businesses, James Hardie had a sophisticated and well-trained section devoted to environmental control and industrial hygiene. The Environment Control and Industrial

Hygiene Unit at the firm's Camellia factory was responsible for monitoring the dust levels at Baryulgil.[32] Information on what were considered acceptable levels of pollution was issued from head office to managers of all branches of Hardie plants including the manager at Baryulgil. Samples of dust levels taken at the mine were sent to Camellia where they were examined by senior technical officers such as Mr J. Winters and Dr S. McCullagh. The annual X-ray programme which was introduced by the company in the late 1960s was coordinated by Dr McCullagh. Decisions involving capital expenditure which came before the Board were also as a matter of course discussed with the Industrial Hygiene Section and the Environmental Control Committee.[33] From the early 1970s there was close consultation between the company and the Division of Occupational Health (located within the New South Wales Health Commission) and the Mines Inspectorate on the use of dust-monitoring techniques and equipment. The degree of expertise held by the company was in advance of those departments and the flow of information was very much in favour of the government bodies. The Mines Inspectorate was in fact largely dependent upon James Hardie in keeping abreast of knowledge about the monitoring and control of asbestos fibre in the workplace.

During the life of the Baryulgil mine various state government departments shared responsibility for the inspection and regulation of the mine's operation. There was at no time, however, comprehensive regulations which covered industrial hygiene in the asbestos industry. The first formal regulations in New South Wales which applied to an asbestos mine came into operation in 1964, setting a specific statutory limit on the amount of airborne dust permissible. These regulations were based on a Victorian standard which had been in operation for some twenty years. The Victorian legislation was not directed specifically to the asbestos industry but included asbestos as but one of a number of substances provoking respiratory illness. That legislation was for the time comprehensive and refers to the provision of exhaust ventilation and respirators, the vacuum-cleaning of workroom surfaces, the use of wet brushes in sweeping floors and benches, and the instruction of workers on hazards. There is also reference to the provision of showers and locker and lunchroom facilities. The Victorian code allowed for the periodic testing of air in the workroom and the routine medical examination every six months of all exposed workers. The legislation makes specific reference to the need for careful recording of such information placing the responsibility upon the owner to "insist on complete and continuous records of these findings and transfer all affected workers to other work".[34]

The role of the regulatory authorities at Baryulgil is a vital element in the history of the fate of that community. The Mines Inspectorate which had responsibility to ensure that the mine was properly run is now found within the Department of Labour and Industry and the Department of Mines. The Mines Inspection Act of 1902 stipulates that a quarry such as that of Baryulgil should come within the terms of the Act, thereby giving the Mines Inspectorate the role of licensing both the mine and the appointment of its manager. The Act confers the right of inspection at any time of day or night and there is no suggestion that inspectors should forewarn owners of such visits. Under the Act, it is the duty of inspectors to notify owners of any health hazard and to specify the nature of such a hazard and the measures necessary for its remedy. An inspector is under a legal obligation to issue such notices where necessary. The legislation also contains recognition of the duty of employers to provide the department with information of any known risk. Under the amendment, Section 32(2B), which was introduced in 1967, some twelve years before the mine's closure, the owner is required to notify the Chief Inspector of Mines within seven days of any case of pulmonary disease among its workforce. On paper, at least, the Act appears adequate for the monitoring and regulation of any mining operation in the state.

The earliest surviving record of an inspection at the Baryulgil site is dated 20 November 1953. It is know, however, that inspections were carried out in 1948 and 1952, but the records of these inspections have disappeared. It is also known that the dust counts obtained during these visits were high, and yet the Inspectorate declined to take any action against the manager or owner.[35] It was a further eight years before the Inspectorate saw fit to once again take readings of dust levels. Why such a decision was taken is unknown.

There were in all some ninety visits by department officers to Baryulgil in the period from 1953 to 1981. In general the reports from these inspections give a favourable impression of conditions at the mine. They contain no suggestion that there was any health risk from dust and no warnings were issued either to the manager or to the workforce.

Apart from the Department of Mines, the Division of Occupational Health, located initially within the Department of Health and later within the Health Commission of New South Wales, also carried out dust and fibre counts at Baryulgil. Often these counts were done at the request of the Mines Inspectorate and/or the Dust Diseases Board. One of the division's functions was to provide scientific and technical expertise to the Mines Inspectorate. This led to the anomalous situation in which the Mines Inspectorate had powers to enforce compliance in industrial hygiene but no expertise to monitor risk, whereas the Division of Occupational Health had the expertise but no statutory authority to enforce change. In the case of Baryulgil, the Mines Inspectorate always received copies of any tests carried out by the division. It was from these reports that compliance or noncompliance with the Act was to be judged.[36]

The division was first involved at Baryulgil in 1948 and later took dust counts in 1952 which showed a definite risk to health. These tests were done during the life of the Old Mill, when conditions were at their worst. It is clear from oral evidence that the mine was, by any standards, filthy, and yet the division took no action and there is no record of a warning being issued. It was not until 1970 that the division became regularly involved in inspections of the mine, and from that time its visits were frequent, culminating in the Health Survey of 1977. That survey appears to have been initiated not by the division or by the Mines Inspectorate but by pressure from within the Baryulgil community and from the daily press. The company was itself sceptical of the division's dust readings which invariably showed levels below those obtained by James Hardie's own officers. A Joint Survey was carried out by the division in the company of

James Hardie scientists in August 1971. The Public Interest Advocacy Centre, which tabled a submission before the Baryulgil inquiry, made the following comment on this survey and the reasons government agencies were so inactive in regulating the mine's activities: "It was this type of 'joint survey', based on the artificial conditions which were often orchestrated in the mill for government testing sessions, and initially the superior dust monitoring technology available to the company, that contributed to the relatively passive and misinformed role the government agencies played at Baryulgil."[37]

The final government authority which shared responsibility for conditions at Baryulgil was the State Pollution Control Commission. Although the commission has no specific Act dealing with asbestos waste, both the mine and the mill fell within the ambit of the State Pollution Control Commission and the Clean Waters and Clean Air Acts.

At no time were any of these Acts invoked during the James Hardie period at Baryulgil. No tests were carried out by the commission and no constraints were placed upon the owner or manager. The commission granted the mine a licence which was issued on 28 April 1977 without, as the manager recalls, the commission's agents even visiting the mine site.[38] When an officer from the commission finally visited Baryulgil, he found evidence of a hazard from the tailings which were spread liberally about The Square. Because of the possible health risk to the children at the Baryulgil school, this report was passed on to the state Department of Education.[39] Whatever the failings of the commission at Baryulgil, its inaction should in all fairness be seen in the context of the performance by other government agencies whose involvement was more intimate and stretched over a far longer period of time.

Two types of inspection were carried out at Baryulgil by the Mines Inspectorate. District inspectors visited the site every ten weeks or so to check on the condition of the plant and assess visually the operation of the mine. The measurement of dust levels was done by Special Duty Inspectors whose visits were less frequent. These tests were carried out by officers from the Department of Health on behalf of the Department of Mines.[40]

The company also took its own dust counts and these were done by the local manager and by officers from the firm's Camellia plant. According to the company, when any trouble spot was detected recounts were then taken monthly until the problem had been resolved.[41]

Prior to 1964 there were no statutory limits on asbestos dust levels in New South Wales. In July 1964 a limit was set at 5mp/cf which was, of course, the standard then operating in Victoria and to which James Hardie had for some time adhered voluntarily. On 23 January 1973 this standard was changed from particles of dust to numbers of fibres and from that date the limit was set at 4 fibres/ml. The new technique used in this standard was the membrane filter method, which was welcomed at the time as a revolutionary advance on existing technology. On 2 March 1978 the standard was again changed, this time being lowered to 2 fibres/ml, but the change was not gazetted until May of the following year. It is difficult to compare readings taken over the period because of changes in the units being measured (whether dust particles or individual fibres) and changes in the technology being used in assessing risk. Some readings represent short-term samples, as do those taken by the Mines Inspectorate between 1972 and 1975. After that date, long-term personal sampling was used which involved testing over a four-hour period. The company itself employed both long-term sampling and engineering samples that were intended to identify trouble spots within a plant. By the company's own admission it led government authorities in the adoption of new sampling techniques.[42] However, it is important to remember that, whatever the method used, all techniques achieved little more than an approximation of dust or fibre levels to which a worker is exposed. This is as true of the midget impinger as of the later and more sophisticated membrane filter method that allows for the counting of individual fibres rather than for guesses as to the density of dust. According to the National Health and Medical Research Council (NH & MRC) the membrane filter method is far from an infallible means of monitoring airborne asbestos fibres, and considerable margin for error is inherent in the sampling procedure.[43]

The lack of concern shown by the Mines department in Sydney

about conditions at Baryulgil can be explained by reference to the kinds of reports which the department's own officers made about the mine. An inspection was carried out by the Division of Occupational Health at the request of the Chief Inspector of Mines in March 1960.[44] The report from this visit refers to the dust levels as being "considerably lower" than those found during the previous inspections made in 1948 and 1952.[45] Mention is also made of the installation of a new plant which supposedly had virtually eliminated the hazard that existed in the previous years: "From the result, it is seen that in only one position [in the mill] is the standard concentration exceeded."[46] Despite this optimistic appraisal, the report also refers to the bagging section as being very dirty and the inspector comments on the absence of respirators.[47] At this time the inspectorate was using the unofficial standard of 5 m/cf. Inspections carried out on 11 June 1963 and 24 August 1969 were also favourable, although in another report there is reference to a problem with dust which states: "The atmosphere was visibly dusty in most parts of the mill.[48] The department, however, chose to take no action.[49]

The attitude of the Mines Inspectorate resulted from a number of factors including a lack of expertise. However, there is also a thread of blind optimism which appears again and again in departmental documents from the earliest days of the mine until its closure in 1979. The department's annual report for the year 1946 describes the mine at Baryulgil, giving details on the level of production and the numbers of men employed. The report also refers to the treatment plant which is supposedly "being redesigned and particular attention is to be paid to the suppression of dust".[50] In the following year the annual report refers to "new dust trunking throughout the mill" and to "new huts for mine workers that are in the course of construction".[51] There is no evidence that any such work was carried out, and former miners cannot recall the "new huts" or any huts for that matter which were provided by the company for their use.

The major obstacles faced by the inspectorate were caused by a lack of trained staff. In 1947, for the state of New South Wales as a whole, there were only five superintending inspectors working under the Factories and Shops Act. Among other tasks, these men were responsible for monitoring factories producing asbestos cement sheeting. Like the Mines Inspectorate staff they were chronically overworked.[52] But whatever the demands placed upon officers from the Mines department there is also evidence that they were at times barely competent. An inspection carried out at Baryulgil in August 1972 indicated the presence of a serious problem. The report noted, "All areas except those outside the plant show fibre counts above the statutory 4 fibres per cubic centimetre."[53] And yet no notice was issued to the owners and no action was taken to close the mine or to ensure that the problem was corrected. A similar lack of concern is found in the failure to take action following an inspection carried out in October 1973 which revealed a persistent problem at the mill. The author of the report comments, "The exhaust at the top of the mill continues to emit a constant stream of dust like a dry wood fire . . . the use of hessian bags makes handling of the product a large source of dust."[54]

It is clear from even a cursory reading of the Mines Inspectorate documents that the department made little attempt to regulate or control activities of Asbestos Mines Pty Ltd. At best the inspectors had absolute faith in the goodwill of the manager and owner to correct any fault and to protect the workforce. The attitude taken by the inspectors was explained at some length by Mr Robert Marshall, Chief Inspector of Mines, at his appearance before the House of Representatives inquiry. Mr Marshall explained that it was normal to forewarn the manager of the mine about pending inspections in order to guarantee access to the site and also to maintain goodwill with the personnel involved. He commented, "If we adopt the policy that nobody is to be notified, it is going to create a few hassles."[55] Presumably the mine could be closed on the day of the visit or the inspector may not have been welcome because of some inconvenience his presence may have caused to the management. According to former miners, the mine was always cleaned up for the benefit of the inspector, thereby making the practice of forewarning self-defeating. One officer from the Department of Health who visited the mine recalls that no attempt was made to hide the fact that the mill and its surrounds had been tidied up especially for the visit. Dr Francis remembers, "In the test we did in 1972 when I was there it was obvious that the place had been washed down. The mill, for instance, was wet. The ground was wet. There was no secret made of it. It was quite obvious that it had been hosed down."[56]

At the inquiry, James Hardie sought to justify this practice. According to the company's representative the cleaning up of the mine had no sinister connotations but was just a normal part of "good housekeeping". At a mine site just as in a suburban setting, it is usual to clean up before the arrival of visitors.[57] Furthermore, at no time did the Mines Inspectorate express dissatisfaction with the condition of the mine or mill. The company and the Inspectorate were able to work quite happily in unison and apparently they shared a convergence of interests in keeping the mine open. This convergence of interest is found also in the dependence of the Inspectorate upon the Hardie organization for technical advice. In 1973 a committee was formed within the Mines Inspectorate to introduce a new dust-counting technology. Members of that committee were drawn from the Division of Occupational Health and from the Commonwealth Department of Health. There were also representatives from the Hardie organization.[58] Relying upon the expertise of the company, the committee managed to develop a testing method using the membrane filter which was soon after adopted by the NH & MRC as a standard. The relationship between the company and the department was always cordial.

In defence of its industrial practice at Baryulgil, the Hardie organization has argued that oral evidence from miners about conditions at the mine twenty or twenty-five years ago must be treated with suspicion. In any small community, opinion can be exaggerated and what was an aberration becomes translated in retrospect into an everyday occurrence. A mill prone to one or two days of dust each month is then seen as habitually filthy and every death in the community in the past becomes an asbestos-related death. The company is correct in its claim that evidence from the Department of Mines' files suggest that the mine complied invariably with then current legislation and Hardie's behaved responsibly towards its employees. There is no suggestion that the Inspectorate was unhappy with the way the plant was run, and good relations between the local manager and the inspectors were maintained throughout the life of the mine. This creates a problem of how to reconcile the oral evidence from the community with the evidence from the official sources and those claims made on behalf of the company itself. Either the mine did or did not comply with existing standards of industrial hygiene. The former miners claim it was filthy, as do various independent witnesses who visited Baryulgil at the time. James Hardie claims that the mine was always well maintained and that conditions were adequate. There is, however, a third source of evidence we have which comes in the form of the Hardie Papers. These papers were presented to the inquiry by the Aboriginal Legal Service and they provide the most important documentation about conditions at Baryulgil.

The Hardie Papers are for the most part a collection of inter-house correspondence between the mine manager and head office in Sydney and they cover the period of the managership of Jerry Burke. The papers were supplied to the Aboriginal Legal Service by Mr Burke and present a picture of conditions at the mill which is entirely consistent with the oral evidence from former miners. The Hardie representative at the Inquiry, Mr James Kelso, vacillated on the question of the authenticity of these documents and the company sought to have their contents suppressed. Speaking on the firm's behalf, Mr Kelso argued that if the papers were in fact genuine then they must have been stolen.[59] The documents are in any case incomplete and give a distorted view of conditions. The owner of the documents, Mr Burke, is currently engaged in litigation against the company alleging negligence and breach of statutory duty. He is, therefore, a hostile witness and, according to Mr Kelso, his evidence should be treated with caution. However, in its final report, the House of Representatives inquiry accepted the Hardie Papers as authentic, thereby making them the single most important source of documentation we have about conditions at the mine. The objectivity of the documents cannot be questioned. The authors, all James Hardie employees, never suspected that the contents of these papers would be open to public scrutiny.

The most revealing of these interhouse papers cover the period from 1966 to 1976 when the mine was sold to Woodsreef, but all contain useful information. The first of these papers dated 26 February 1966, is signed by the Chief Draftsman, E. G. Reeve, and refers to a visit he paid to the mine.[60] Reeve comments on the practice of fibre being dropped on the floor of the bagging house where it was then shovelled up and bagged. He makes various suggestions on measures to contain dust in the mill including total enclosure and the introduction of exhaust fans. During Reeve's visit photographs were taken in the dirtiest part of the mill. Reeve explains, "The photographic processor did not print the sock-loading operation presumably regarding it as blank film, which it most certainly is not."[61] The obvious explanation is that the dust about the sock area was so intense that the processor found no subject visible on the print, and therefore assumed the film to be blank. None of the recommendations made in Reeve's report were acted upon.

Document number 89a[62] is a mine manager's report for the fortnight ending 1 April 1969 which suggests that conditions at the mine were poor. In a summary of the dust readings taken for this period the author comments, "The only place [within the mill] which is approaching the tolerable limit is the bagging area, which is 213 pcc."[63] The report is signed by the manager, Jerry Burke. Much the same picture is found in Document 5,[64] which is a summary of the Industrial Hygiene Survey carried out at the mine between 14 and 17 September 1970. The paper is signed by Mr J. Winters, the company's Industrial Hygiene Engineer. He explains that the manager is well aware of the problem of dust at several locations where levels are "alarmingly high". There follows an account of readings taken at eight specific points or stations in and around the mill. Station 5 is cited as recording 245 f/cc and a number of other readings are also well above the prevailing statutory limit. The report also mentions that cleaning of the mill floor was being done by hand with a broom because the vacuum cleaner was being repaired. Winters recommends the need for the socks in the dust house to be shaken mechanically rather than by hand as was then current practice.

In November of the same year an interhouse letter from Dr

S. F. McCullagh,[65] the company's senior medical officer, to the Baryulgil manager gives some indication of the reliability of the dust readings taken by the Health department. Dr McCullagh mentions a survey carried out by officers from that department in 1969 which showed only one site at which the dust level was above the statutory limit. He comments that head office was disinclined to accept these figures and that at the time Mr Winters found only two stations out of nine tested where the levels were satisfactory. The same disregard for the readings done for the Mines Inspectorate is found in Document 20, dated 21 February 1974.[66] On this occasion, Dr McCullagh remarks, "The asbestos in the air levels recorded by the Inspector are lower than may correctly be found at the Mine." The inspectorate's error is due, he believes, to a lack of expertise, and also to the fact that over 150 millimetres of rain had fallen in the area just before the testing was carried out.

In the period from February 1972 until October 1976 which is covered in Documents 48 to 63 there are ample warnings that conditions at Baryulgil were anything but satisfactory. Some of the interhouse reports are such that it is difficult to understand how the warnings could have been ignored. One report dated 7 February 1972[67] contains the following description of the mine site: "Nevertheless, billowing clouds of fibre could be seen coming from this building [the mill] and Mr Burke tells me he has, on occasions seen such clouds from distances of several miles." In a second report made in the same month,[68] Dr McCullagh comments that despite "some marginal improvements there is little change and the picture remains gloomy". But perhaps the most disturbing aspect of the Hardie Papers is not the indifference shown towards the persistently high dust counts but the levity with which the counts are treated. In a note from Mr Winters to Jerry Burke dated 21 September 1976[69] there is reference to the need to provide an airline respirator for the men in one particularly dusty station and to the generally high readings from the previous month's inspection. The note concludes with the comment to Burke regarding the pending sale of the mill to Woodsreef: "From the results it looks as though you intend to go out on a *high* note." Although the comment is casually made and appears at the bottom of the memo, it does suggest that the presence of dust in the mill was not taken seriously at head office. Such an attitude helps to explain the lack of action over a period of many years to correct obvious flaws in the plant and to curtail the high dust readings which the Hardie Papers document.

The difference between the conditions in the mine described in the Hardie Papers and the evidence from the Mines Inspectorate is due in part to the practice of cleaning up the site before inspections. According to the oral evidence presented at the inquiry, clean-ups and the deliberate slowing down of the plant were done routinely prior to all inspections, whether by government departments or by officers from James Hardie's head office. Burke recalls that it was usual to clean up the mill "out of courtesy to the person who was coming".[70] As manager he denies having issued orders for the clean-up or for the operator to slow down the plant but he did not deny that such things happened. Bill Hindle, who also gave evidence before the inquiry, remembered that during his twenty-five years at Baryulgil he never saw an inspector from the Mines or Health departments make a spot check at the mine. There was always one or two days' notice given.[71] Hindle also recalled that during such visits the plant would be slowed up and the rate of feed turned down to "just a trickle". This, of course, reduced the amount of dust in the mill. Like Burke, Hindle could not recall anyone issuing the order for the slow-down: it was just a matter of habit.[72] Even if the forewarning had not been given it was difficult for anyone to arrive unnoticed at such an isolated town as Baryulgil. According to Hindle the only person who managed to do so was the "sly grog bloke" who for many years sold alcohol illegally at The Square.

Former miner Neil Walker recalls the way in which the clean-ups would take place. "First of all he [the manager] would get all the hoses and jackhammers down in the quarry and he would run it up to the mill. They used to hook it up to the water pipe. It used to run from the quarry up to the mill. They would pump the air through that way. They would go around with the air hoses blowing to the ground all the dust that used to lie on the beams on top, and once it hit the ground they swept it up."[73] This practice was not questioned by the men and they saw nothing unusual in seeking to present the best possible face to visitors. What is anomalous about the practice is that it was also carried out for the benefit of visitors from James Hardie who apparently failed to notice the feeble rate at which the plant was operating. Neil Walker believes that Frank Page, who was a regular visitor from Sydney, must have known the plant was being slowed down for his benefit.[74] Page was a trained engineer and he was also aware of the production quotas for the mine, which could hardly have been met while the plant was being run at half-speed. Walker believes that in the case of Page the slow-downs were done in order to avoid embarrassment. If the plant were run normally, Page would have been forced to make a decision as to the future of the mill, as its design was clearly inadequate for the demands being made upon it.

James Hardie has denied that orders were issued by the company for the slowing down of the plant.[75] It acknowledges that there may have been some effort made to clean up the mill, but such behaviour involved nothing more than a matter of etiquette. According to Mr Kelso, "Your wife puts all the dishes out of the sink into the dishwasher before somebody comes and I believe that would happen at Baryulgil as well."[76] Presumably, neither Mr Kelso nor the company he represents perceives the practice of cleaning up the mill as raising ethical questions.

In a dusty mill such as that at Baryulgil it is essential that workers be given the protection afforded by respirators. The Victorian legislation of 1945 and the British legislation of 1931 both emphasize the need for such protective equipment whenever dust cannot be suppressed. According to the company, instructions were given by the local manager to all employees on the use of breathing apparatus. At the inquiry Mr Kelso explained that there was some resistance by the workforce to use face masks and it was a matter of getting the men to "do the right thing" rather than of taking disciplinary action.[77] As far as the company was concerned the equipment was provided and it was used when necessary. The responsibility for ensuring its use lay with the local manager.[78] This, however, gives little comfort as to the availability and use of such equipment during the early days of the mine, and evidence from former miners suggests that respirators were rarely worn. The masks were faced with re-placeable pads which, in the heavy dust of the mill, would soon clog, making breathing difficult. In such heat, men doing heavy work could not wear the equipment.[79] Speaking in the company's defence before the inquiry, Mr Kelso quoted from a letter written by a female visitor to the mine. The letter in part states, "However, he [the manager] insists that safety helmets are worn by everyone. Even though operations are performed by shoeless workers, Jerry accepts this as he feels his Aboriginals are fleet-footed enough to jump away from danger."[80] Perhaps Mr Kelso did not intend to convey the incongruous picture of bare-footed men wearing hard hats and respirators.

Whatever its failings over dust control and the provision of safety equipment, the company did, from 1969, participate in regular health checks of the mine staff which were carried out by the Health department. The surveys involved spirometry tests and X-rays and the results were made available to the company. From 1973 to 1976, half the workforce was tested annually. According to James Hardie, the New South Wales Department of Public Health carried out surveys at Baryulgil on six separate occasions from 1948 to 1969. The results from the early tests are, however, not available and it is not known if there were any warnings in these medical surveys as to the existence of disease at the mine.

One important, if obscure, issue debated during the inquiry concerns the presence of used asbestos bags at the mine site. There have been two recorded cases of mesothelioma among the Baryulgil community. Bill Hindle, the former fitter, died from the disease in December 1984, and a woman from The Square is also known to have suffered from this rare form of cancer before her death. The former manager, Arthur Allen, and his wife are both believed to have died from cancer after they left the mine in 1966. In this context the question as to the origins of recycled bags becomes significant.

The fibre from Baryulgil was packed in hessian bags and trucked by Keech Transport to the railway at Grafton. This work was later performed by Farquhar Transport. It is known that one of Keech's former employees has brought legal action against James Hardie for injury caused by asbestos leaking from the bags. According to the Hardie Papers the hessian bags were a constant problem, spilling fibre and dust and making the packing and removal of the product hazardous. The bags were handled manually throughout the process of filling, sewing, storage and transport. According to the company's own records the bags would stretch, thereby releasing dust and fibre over the workmen.[81] Many of the bags used at Baryulgil were recycled and contained residues of asbestos fibre from other mines. The late Bill Hindle recalled that the bags displayed the initials EGNAP which referred to a South African producer. There were also bags from Wittenoom.[82] The bags came in three sizes: 50-, 75- and 100-pound units. They would arrive at the mine packed in lots of one hundred. Because of the poverty of the Baryulgil community, the bags often found their way into the miner's homes where they were used for a variety of purposes. The former miners still remember the long blue and brown fibres which they would find at the bottom of the bags. Neil Walker and Harrington both recall the residues which displayed fibres far longer than those mined at Baryulgil.[83]

The company has disputed these claims and also denies that 100-pound bags were ever present at the site.[84] It is adamant that no bags containing residues of blue asbestos were used at Baryulgil. Clearly, one of the parties is mistaken. If blue asbestos from recycled bags was introduced into the community it would provide an explanation for the cases of mesothelioma which have occurred at The Square. If not, then the cause of these deaths remains a mystery.

James Hardie's involvement at Baryulgil came to an end on 23 September 1976 when it sold all shares in Asbestos Mines Pty Ltd. The buyer, Woodsreef, was interested not in the Baryulgil site but in the mining tenements in the area to which it already held adjacent leases. At the time of purchase, Woodsreef agreed to continue working Baryulgil until at least June of the following year in order to guarantee employment for the local workforce. The mine, however, continued in operation beyond that date and closed finally on 24 April 1979. At the time of sale in 1976, Jerry Burke stayed on as manager as did ten or twelve of the original miners who formed the reduced workforce used by Woodsreef.[85] Although it had worked the mine for only a brief period, Woodsreef carried out extensive restorative work after its closure at the request of the Department of Mineral Resources. This work included grading the site, fencing the pit, the removal of the old mine buildings and the planting of trees on the tailings heap.

The Baryulgil mine was only a small concern in comparison with the chrysotile mine operated by Woodsreef at Barraba. Mining began at Barraba in 1971 with projections for the processing of 2 million tonnes of ore within three years. The company was owned by Canadian interests and the product was to be sold to the expanding Japanese market. According to the former miners, the change of ownership at Baryulgil did nothing to improve conditions in the mill.[86] There was no capital expenditure at the site which is not surprising given Woodsreef's motives in purchasing the lease from James Hardie.

The closure of the mine in 1979 ended a period of full employment for the people of Baryulgil. For nearly two generations Asbestos Mines Pty Ltd had protected the community from the unemployment which was the fate of most Aboriginal people in New South Wales. The mine's closure also ended the exposure of the men, women and children of Baryulgil to intolerably high levels of asbestos fibre which the mill, from the first days of its operation in 1944, released freely over The Square.

7

Baryulgil: The People

It is tempting to view events at Baryulgil as an example of the ruthless exploitation of an Aboriginal labour force by a major Australian company. Each case of asbestosis among the miners would thereby be seen as proof of a rampant corporate greed.

During the inquiry, this kind of interpretation lay behind much of the argument presented by the Aboriginal Legal Service and by the Public Interest Advocacy Centre. There is no evidence, however, that James Hardie made money from the misery which asbestos-related disease has brought to that community. The mine was never an important part of the Hardie empire and returns from the operation were always marginal.

The mine was purchased originally by Hardie in order to guarantee a local supply of fibre, but for only a brief period did Asbestos Mines Pty Ltd operate successfully. In the years 1969 and 1970 the company made a profit. For the rest of the years of its existence it did not.[1] By its own admission, the parent company kept the mine open out of a felt obligation as the only local employer. The company was aware that, if the mine were closed, the workforce would be faced with permanent unemployment and this was "one of the reasons why the mine was persevered with in a time of falling profits".[2] In such a context, Woodsreef was an attractive purchaser for James Hardie when it sold the mine in 1976, because Woodsreef promised to keep mining the site.

Within the wider company structure, Baryulgil was insignificant. At its peak the mine produced a mere 1 per cent of Hardie's total fibre needs.[3] The product from Baryulgil was coarse and

less regular than imported fibres and the yield from the ore body
was not particularly good. The parent company was for some
time undecided as to how long it would keep the mine open, and
it appears that Baryulgil managed to survive more by chance than
by design. Given the marginal profitability of the mine and the
unimportance of its product, it is surprising that it lasted so long.

The ALS claimed at the inquiry that Hardie retained the Baryulgil mine in order, as a local producer, to have the right of representation before the Tariff Board in deciding policy on imported fibre and asbestos products. The ALS, however, has produced no evidence in support of its claim. The James Hardie company explains its operation of Baryulgil on the grounds of benevolence and, supposedly, it kept the mine operating for the sake
of the community and not out of any self-interest. This claim
can be judged against the behaviour of Hardie as an employer at
Baryulgil and by examining the actual living conditions at The
Square. The obligations which fell upon Hardie were the same as
those which are accepted by any employer operating a company
town.

Throughout the life of the mine there was little or no trade
union presence. In the 1950s, there was no union representative,
and there was no trade union membership. In the following decade, many of the men joined the Australian Workers Union, but
the union did not take an active role in the question of working
conditions. It appears that the local AWU organizer was more
interested in the collection of union dues and the payment of
wages, than with issues of industrial hygiene. Rodney MacBeth,
the former AWU organizer at Baryulgil, recalls that his job at
the mine was to secure enrolments and to attend to any disputes
between workers and management. If he had received complaints
about the dust he would have discussed the issue with the manager and then placed a report with central office.[4] MacBeth first
visited Baryulgil in 1974, and from that time until its closure in
1979 he visited the mine regularly. He cannot recall if the AWU
ever complained officially to James Hardie about industrial
hygiene. He does remember, however, that Baryulgil was the only asbestos mine he visited, and that he knew nothing about the
possible dangers from airborne fibre. When questioned at the in-

quiry as to whether he worried about the dust, MacBeth replied: "Yes, to a certain extent, but when you approached the place on a still day there was always a haze about. To be quite candid the same thing applied to cement works."[5]

Donald Wilson, a former labourer at the mine, was the local AWU shop steward. He cannot recall that the union ever raised the question of a health hazard, and he cannot remember the AWU seeking to inform the miners about precautions they should take in gaining protection from the dust.[6] Wilson made personal representations to the manager about conditions but, apparently, he did not see fit to raise the matter with the union. He also recalled that respirators were not generally available, that the washing facilities were inadequate, and that the mill was slowed down for the benefit of inspectors.[7]

Wages paid at Baryulgil came under the Metaliferous Miners (Open Quarry) Award and the company never paid more than the award rate. Neil Walker, who worked as foreman, did not receive an above-award payment for his level of responsibility, a payment to which he was entitled.[8] In general, company policy was to pay the bare minimum. The quarry was worked in a series of 7-metre benches. This method was inconvenient, and a more suitable approach would have employed deeper benching. If such a method had been used, however, the miners would have been entitled to "depth money". Although the quarry pit was over 30 metres deep, "depth money" was never paid during Walker's years at the mine. Walker was sacked shortly after returning from long-service leave. On the day of his dismissal, the men under him had completed their quota for the shift and were sitting about talking and smoking. Walker would not tell the men to return to their jobs: they were his friends and neighbours and they had completed their day's work. The manager, Jerry Burke, dismissed him on the spot for allowing the men to loaf, and he refused to reverse his decision which Burke said was a matter of discipline. Walker, subsequently, wrote to Frank Page, a senior Hardie official in Sydney, who merely confirmed that the dismissal notice would stand. Walker believes he lost his job because he took long-service leave and that the same thing happened to John Mundine. The only other worker to take such leave

was Cyril Mundine who afterwards did not return to the mine.
Bill Hindle, the fitter at the mine, did not take leave and neither
did Greville Torrens, although both were entitled to do so.[9] In
Walker's view the company was always miserly, and there is
some evidence to support his allegation.

Warwick Sinclair, a former claims officer with the AWU, remembers an incident at the mine during the mid 1960s involving
the payment of under-award wages.[10] An AWU district organizer
had come across an Aboriginal community living in harsh conditions at Baryulgil. The miners were being grossly underpaid and
they were afraid to join the union. The AWU decided to take action on the miner's behalf even though they were not members.
Sinclair visited the James Hardie head office in Sydney where,
he recalls, he received "the Duchess Treatment". He met a number of senior personnel who were anxious to avoid trouble over
the issue. Sinclair remembers that the wage records were dis-
organized and the Hardie executive sought to lay the blame upon
a junior clerk. After some consultation Hardie sent the AWU a
cheque for $3,000, which was back-pay for thirty men for a
period of twelve months. Although underpayment had been going on for far longer, it was not possible at that time to seek
compensation for more than a one-year period. The union made
out cheques for each of the men involved and George Duncan,
the local AWU organizer, subsequently enrolled all but two of
the miners in the union. Some weeks after the settlement, Sinclair
received a telephone call from the wife of the Hardie pay clerk.
She told Sinclair that the company had pressured her husband to
such an extent that he had been admitted to the Ryde Psychiatric
Hospital. Sinclair believes the clerk was being used as a scape-
goat and had collapsed under the strain. He cannot remember
what eventually happened to the man, but he has never forgot-
ten the conditions at Baryulgil: the heat, the dust and the "inde-
scribable poverty" he saw there.

The underpayment of wages at Baryulgil may have been due to
an oversight, and Sinclair certainly tells his story in a most flam-
boyant manner. Even so, this episode confounds the company's
claims to benevolence in running the mine and suggests, at the
least, a lack of concern for the workforce. What is most disturb-

ing about Sinclair's account is his reference to the appalling poverty he saw at The Square.

James Hardie claims that the mine was vital to the welfare of the community, and that the illness found at The Square is simply characteristic of all New South Wales Aboriginal settlements.[11] The company has also argued, quite correctly, that today the people of Baryulgil are far better off in terms of housing than are other Aborigines from that district.[12] This, of course, is due to the expenditure of several million dollars of public money. It is not the result of any action taken by James Hardie. The company is also adamant that there is no particular health problem among the ex-miners, and that both mortality and morbidity rates are much the same as those in other Aboriginal communities. This, of course, hardly gives a flattering picture of life at The Square in what was from 1944 to 1976 a company town run by one of Australia's largest corporations.

The houses at The Square were constructed, not by the company, but by the people themselves. All kinds of discarded materials were used and the dirt floors were usually made from ants nests which were crushed and wetted to form a rock-hard surface. Squashed tins were used for the walls and discarded asbestos bags were pasted up to form a wall lining. During the 1950s a typical house consisted of one large room with a bed, a table and a kerosene lamp. There was no electricity, no running water and no sewage. Sewage improved gradually during the 1960s, but it was never satisfactory. In all houses, recycled asbestos bags were used as a building and furnishing material. The women would wash the bags and fashion them into sheets and floor coverings and as wall insulation to keep out the wind and rain.[13] One woman sewed several 100-pound bags together to make a bedspread.

In the late 1950s several families at The Square purchased wood from Tabulum to make better houses. They also obtained discarded fibro-cement sheets from the mine, which were carted in the empty trucks from Grafton. With hard work and an assortment of materials the housing was gradually improved. For reasons known only to James Hardie, the company declined to become involved in helping the community in this work.

Water was always a problem at Baryulgil. The creek, which lies one mile from The Square towards the mine, would alternately flood or dry up. In each case, the absence of a reliable water supply made life difficult, especially for the women. Mick Mullins, a local white, would sell 44-gallon drums of water for £2 each. Pauline Gordon remembers using as many as four of these drums each week.[14] The water was used for drinking, cooking and for bathing children. Most washing was done at the creek, and the women would spend a whole day with the children heating up water in 44-gallon drums over an open fire. Apart from water, wood was also scarce and, on weekends, Mullins sold firewood from the back of his truck. Because of the high cost of living and the low wages paid, life at Baryulgil was difficult. There were no cars in the community until the mid-1960s, and the dirt road to Grafton was unreliable and at times impassable, thereby further isolating the community from the outside world. Food was ordered from the general store and there was a mail-truck delivery from Tabulum once a week. The daily food was supplemented by game, such as kangaroos, porcupines and eels.

Throughout the life of the mine, infant mortality at The Square was high. The housing was poor and there was no medical care. For many years a local woman, Mrs Lucy Daley, acted as a midwife. The company which ran the mine neglected to provide even the most basic conveniences, such as a water pump which, for a small investment, would have made life so much easier for the women and their families. Asbestos Mines Pty Ltd's one concession was a Christmas party each year. The men would receive two bottles of beer and there were sandwiches and presents for the children. The presents were paid for by the men themselves. The people at Baryulgil were different from other Aboriginal communities in having regular employment, but this did not mean that their conditions of life were much better.

There was less than a kilometre separating the mine from The Square, and a constant cloud of dust and fibre from the mill was blown over the houses. The miners returned home at the end of each day in their work clothes which were saturated with fibre. It was also common for the women to visit the mine site to bring the men their lunch. The tailing heap adjacent to the mine was a playground for the children, who would slide down its sides and search for birds' nests in the rubble. But the most serious source of contamination at The Square came from the use of tailings as a surfacing material. The district has a high rainfall and the ground tends to become soggy. From the early days of the mine it was common practice to spread tailings about The Square. Six or eight times each year this would be done from the back of a tip truck. According to ALS estimates, in any one year more than 1,500 tonnes of the material would be spread in this way.[15]

Tailings were also used by the Copmanhurst Shire Council,[16] and on the playing areas of the Baryulgil public school, which was attended by most of the children from The Square. The children would play in the dust and would fashion jump pits from the fibre-rich gravel. This practice continued until 1977, even though knowledge about the environmental hazards of asbestos is clearly suggested in Wagner's pioneering work carried out in South Africa some twenty years earlier. According to the ALS, responsibility for the use of tailings at The Square rests with the company. The material was supplied from the Hardie mine and it was transported in the company truck. Furthermore the company was the only party which must have known that the tailings were potentially hazardous.[17] Whoever is responsible, the practice of using tailings as a surfacing material was a weekly occurrence.

Jerry Burke, who lived at the mine for more than twenty-one years, remembers asbestos fibre as an inescapable part of life: "We got asbestos fibre on the roof all the time and we were drinking it in our tank water."[18] At Baryulgil the prevailing wind blows from the south to the north, taking dust and fibre from the mine out over the town. According to Burke, "The mill generated a pall of dust, you could see it in the sun of an afternoon the dust going out towards the north west. This took it over my house and then over towards The Square and all towards Yulgilbar station area."[19] For most of his years at the mine, Burke's wife suffered from a bronchial disorder and in 1983 one of his sons, who grew up at The Square, was found to have a slight patch on his lungs.

Burke's memory about conditions is supported by Charles

Sheather, a former miner. Sheather recalls that one of the managers used to let his children swim in the quarry pump area. This was a pool into which the water from the quarry was emptied with the overflow from the pool running past the manager's house. On one occasion Sheather let his own son swim in the pool but he soon forbade the practice after seeing the amount of fibre which collected on the boy's swimming costume.[20] According to Mrs Gordon, who has spent all but a few years of her life at Baryulgil, the public school itself was built on 2 metres of tailings which were used to level the ground. She also remembers the amount of dust which tailings, used as a road surfacing material, would create. Dust at The Square was stirred up by cars and when they drove past "you are coughing and choking and you have to block your nose. You have to wait for it to settle and then you can breathe again."[21] And, for the women, there were also other dangers. Before washing, the women used to beat the men's clothes on a tree to free them from the fibre which stuck tenaciously to the fabric. This was carried out two or three times each week, and no precautions were taken to avoid inhaling the fibre set free as the clothes were struck against the tree trunk.

Neil Walker is one of the residents who used tailings on his lawn, where he would spread it about several inches thick. The tailings levelled the surface and they also encouraged a lush cover of grass. According to oral evidence from miners and their wives, there were innumerable ways that Baryulgil residents would come into contact with airborne asbestos. It was not possible to live at Baryulgil and escape exposure. According to James Hardie, evidence from residents about asbestos dust at The Square is misleading, as they habitually confuse serpentine dust with asbestos fibre.[22] The company does not deny that dust did drift out over the town, but it was essentially harmless dust from the host rock and, presumably, it did not contain asbestos. In similar fashion the company rejects claims, made by the ALS, that tailings used by the Copmanhurst Shire and at the Buryulgil school presented a hazard. Hardie is certain that the tailings consisted of serpentine dust with only a tiny amount of asbestos. That presence is put by the company at between 2 and 5 per cent — a level which presents no risk to health.[23] Whatever the effect of the tailings, the James Hardie organization is, by its own account, in any case not culpable because the material was used without its approval or knowledge.[24] The practice took place as the result of an agreement between the local mine manager and the Baryulgil community. Consequently, any injury is thereby the fault of the manager, who acted on his own initiative. Mr Kelso explained at the inquiry that James Hardie had taken samples of the tailings, which revealed only a trivial presence of fibre. One batch examined showed less than one-third of one per cent of asbestos, while a second revealed only a marginally higher figure.[25] Hardie is convinced that there is no risk to any one living at The Square now or in the past. This opinion, however, is not shared by the Aboriginal National Conference (ANC), which also made a submission before the inquiry.[26]

The ANC believes there is a health hazard at The Square for all residents and, particularly, for the children. In its submission the ANC points out quite correctly that the original mining operation did not extract all the fibre from the host rock, thereby leaving large quantities of asbestos which are visible in the tailings. The presence of tailings was obvious to ANC officers in parts of the school yard and around the houses and even on the roads leading to The Square. This judgment is supported by a scientific study commissioned by the ALS. The study, which was carried out by Mr K. Cross from the Department of Geology, University of New South Wales, in October 1980, revealed ample evidence of airborne fibre. In his report, Mr Cross comments, "There is no doubt that the residents of Baryulgil are currently being exposed to highly undesirable levels of asbestos dust."[27] He also points out that it is difficult and expensive to remove or to cover such tailings and that, even if buried, with the passage of time and weathering the problem would only re-emerge in the years to come. It is important to remember that the work which has been carried out at Baryulgil to cover the tailings was not initiated nor has it been paid for by the company which originally created the pollution, namely Asbestos Mines Pty Ltd. The cost of the work has been borne by the people of New South Wales.

The First Evidence

Awareness within the community of the dangers posed by asbestos fibre at Baryulgil grew slowly. No warning was given to the workforce by the manager or the owners of the mine, and the Mines Inspector took no care in ensuring that the miners were aware of the potential risk to their health. Knowledge in the community grew in response to the deaths of a number of prominent men who had spent their work lives in the mine. The first of these deaths was that of Cyril Mundine who, for many years, served as foreman in the quarry. Mundine was one of seven brothers, all of whom, with one exception, are now dead. The sole survivor is the one brother who did not work in the mine. Cyril Mundine was employed by Asbestos Mines Pty Ltd from 1944 to 1965. Most of this time was spent using a jackhammer, drilling holes for blasting the ore and breaking down large blocks of ore into workable pieces. Claude Mundine also worked in the quarry, as did Richie Mundine, who spent some time in the bagging section. Jerry Burke recalls that there was some talk that Richie died from asbestosis.[28]

Cyril Mundine died of asbestosis and the Dust Disease Board (DDB) awarded his widow $3,000 in compensation. Unfortunately, prior to his death, Mundine had moved to live in Sydney, thereby depriving the community of the warning his death should have signified. In the case of the other brothers, evidence from people who remember the Mundines suggest that, for most, their final illness was consistent with the presence of asbestos-related disease. According to the ALS, in all there have been as many as seventy such deaths in the community.[29] None of these people was diagnosed at the time as suffering from asbestosis and, according to one witness, most were believed to have died from chest problems and from heart disease.[30] This, in itself, is not surprising, given the difficulty even for a specialist in diagnosing asbestosis.

The oral tradition depicting asbestos-related disease at Baryulgil is strongly disputed by James Hardie which will not even acknowledge that asbestos played a major role in the death of Cyril Mundine.[31] Speaking on the company's behalf, Mr Kelso tabled a copy of Cyril Mundine's death certificate before the inquiry. This document shows the cause of death as pulmonary oedema with gross congestive cardiac failure.[32] Furthermore, the death certificates of the other Mundine brothers make no reference to asbestosis or other asbestos-related illness as the primary cause of death. In the absence of an autopsy, the worth of death certificates as a guide to the presence of asbestosis is dubious. At the time of his death, Mundine was in receipt of a pension for asbestosis from the DDB. If that is not reflected on his death certificate, the reason may well lie with the attending physician who may not have been aware of Mundine's medical history. The certificate itself is not proof that he died from another illness.

The second death which aroused awareness of the dangers at the mine was that of Andrew Donnelly who died on 17 June 1977. Donnelly was a member of one of the original Baryulgil families, and he had worked in the mine for twenty-eight years. Donnelly played competitive football until his early forties and he was very fit, neither drinking nor smoking. He had never taken a day off from work because of illness, and he was looked upon as one of the leaders of the community. He was, in the words of Jerry Burke, "a totally reliable worker". During the weekend prior to his death, Donnelly worked overtime cleaning machinery and, when he arrived for work on the Monday, he was ill. He was taken to Grafton Base Hospital late that evening, where he died two days later. The community believes he died because of his employment in the mine and, at the inquiry, the ALS claimed that his death was due specifically to asbestosis. The cause of death cited on his death certificate, however, is given as lobar pneumonia.[33] With the agreement of his widow, a post-mortem was performed on the body, and this led to a rather different verdict. Dr K. Murray's diagnosis was one of asbestosis with accelerated hypertension and viral pneumonia.[34] A second document on the cause of death, signed this time by Dr R. J. Grobius of the Grafton Base Hospital, gives asbestosis as the primary cause of death referring, specifically, to gross disease.[35] The contradiction between the causes of death, cited on the death certificate and the judgments made by Drs Murray and Grobius is not unusual. Death certificates are notoriously unreliable.

Mr Kelso, the James Hardie representative at the inquiry, laboured the issue of Donnelly's death at some length. He pointed out, quite correctly, that the DDB had rejected a claim for compensation by Donnelly's widow because the death certificate did not give asbestosis as the primary cause of death. The DDB chose to ignore the evidence from Murray's autopsy, thereby compounding the injustice already suffered by the Donnelly family. That decision appears to have little merit other than adherence to a formal procedure, adopted by the Board, apparently, for its own convenience. That decision, however, does not disprove the claim that Donnelly died as a result of his employment by Asbestos Mines Pty Ltd.

Mr Kelso also explained to the House of Representatives Committee that Andrew Donnelly was, in any case, fully aware of the dangers of working with asbestos fibre. Mr Kelso went on to comment that this awareness "was reflected on more than one occasion when he told the company's industrial hygiene officer: 'It will not hurt you.' He [Donnelly] meant asbestos. This statement clearly indicates that he had, in fact, been advised of the potential health hazard associated with asbestos and was apparently inclined to discount that advice."[36] Presumably, in the company's view, if Andrew Donnelly did die from asbestosis, it was his own fault. Unfortunately, Mr Kelso failed to reveal what the company's hygiene officer had, in turn, said to Donnelly.

Donnelly's death did at least have the effect of alerting the community and a number of government authorities to the immediate danger at Baryulgil. This process was helped by the work of an ABC journalist, Mathew Peacock, who authored a series of radio programmes in July and September 1977 devoted to the subject of asbestos. The programme aired on 30 September featured Baryulgil, and Peacock spent some time at The Square, speaking to the people about the dangers they faced. The Division of Occupational Health conducted a survey into the health of the community in that same year, but it is Peacock who is credited by the people of Baryulgil with warning them for the first time about asbestos fibre. The State Pollution Control Commission was less enthusiastic than the Division of Occupa-

Baryulgil

(All photographs supplied by Aboriginal Legal Service, Redfern, unless noted otherwise)

1. The mill at Baryulgil, which is a small Aboriginal town some two hours' drive northwest from Grafton in northern New South Wales. The mine was always a small operation, employing at its peak less than fifty men. It was a unique community, however, in that the workforce at the mine was recruited almost entirely from among the local Aboriginal communities.

2. Part of the mine.

3. During the first years of the mine's operation during World War II, the host rock and tailings were carted by horse-drawn skip from the mine site. This method was used until the end of the 1940s when trucks were introduced.

4. A miner using a jack hammer to break down ore into workable pieces in the quarry. The miner is wearing a hard hat, which were not introduced into the quarry until more than ten years after the mine reopened in 1944.

5. The bagging section at the mine, which had been partly demolished when this photograph was taken. This was one of the dustiest sections in the plant, and readings taken by James Hardie officials invariably recorded excessive levels of dust.

6. The mine truck was used for various purposes at the mine and mill. The equipment was always rudimentary, and it was serviced by the Aboriginal workers, who also helped to build the new mill which came into operation in 1958. (Photograph by Mr and Mrs Ken Gordon, Baryulgil)

7. The asbestos ore at Baryulgil was mined in an open quarry with drills and the use of high explosives which were used to separate the ore from the quarry face. (Photograph by Mr and Mrs Ken Gordon, Baryulgil)

8. The mine in 1956. It is not certain whether the opaque quality of the picture is due to faulty processing or the dust being generated by the mill. The latter, however, would be consistent with testimony from former miners who remember the mill as being filthy.

9. The Baryulgil mine in the late 1960s. In the foreground is evidence of the asbestos tailings which were used regularly to dress the roadways in and out of the mill area.

10. The mill which was demolished by Woodsreef after the closure of the mine in 1979. Today there is little evidence of the mill as the rubble has been buried over with the tailings from the original mine.

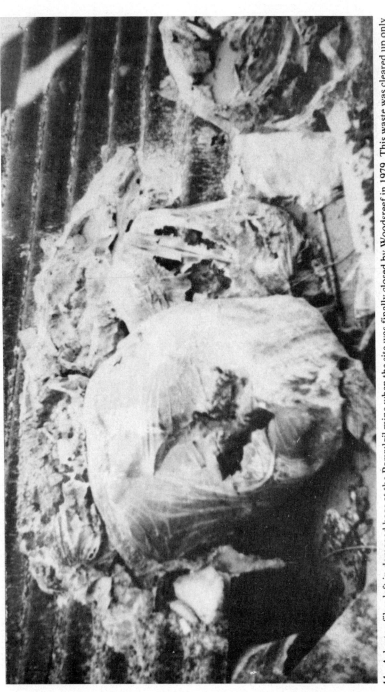

11. Asbestos fibre left in damaged bags at the Baryulgil mine when the site was finally closed by Woodsreef in 1979. This waste was cleared up only after protests from residents.

12. The mine showing packed bags of fibre waiting to be trucked to the railhead at Grafton. From there the asbestos was shipped to Brisbane where it was used by James Hardie in asbestos cement manufacture.

13. The tailings heap adjacent to The Square.

14 (a), (b), (c) and (d). The tailings heap less than one kilometre from The Square. The heap contains tens of thousands of tons of asbestos-rich tailings. The high rainfall has seriously eroded the tailings which are carried by water into the nearby creek and the prevailing winds carry airborne fibre over The Square. Note the meagre attempts at reafforestation carried out belatedly by Woodsreef which operated the mine for the final three years of its life. The planting of trees has not stabilized the tailings and they still represent a danger to the health of the community.

15. A group of Baryulgil residents including Andrew Donnelly (left). It was Donnelly's death in 1977 that alerted the community to the dangers of working with asbestos. Donnelly's legs are covered in fibre as there were no adequate showering facilities or work clothes provided by the mining company.

16. This photograph, taken in the early 1970s, shows two Baryulgil children playing in an area which has been top-dressed with asbestos-rich tailings. The people in the community are adamant that they had no idea about the danger of inhaling asbestos fibre until after the death of Andrew Donnelly in 1977.

tional Health, and its involvement was confined to issuing a licence to the mine. The commission's officers did not visit the mine site, and the licence was signed by the commission's representatives in a meeting with the manager, Jerry Burke, at a town some 40 kilometres from Baryulgil.

The behaviour of the company and of the government departments involved at Baryulgil should not be judged against the deaths of Cyril Mundine and Andrew Donnelly as the first-known cases of disease. There were clear medical warnings much earlier that the mine presented a health hazard. Evidence tendered before the inquiry by the ALS documents medical checks on Baryulgil miners in 1949, that is, barely five years after James Hardie became involved at the site. The first of these documents[37] refers to an X-ray report on a miner named Preece who was examined in January of that year. The report notes that areas of Preece's X-ray are consistent with the appearance of fibrosis, suggesting that the man was in danger of developing serious lung disease. The second document is also a radiology report, this time by a Dr Pooks, on Preece and Harry Mundine. In the case of Preece the physician found what "may be pleural thickening of both lungs". In the case of Mundine he concluded: "Only asbestosis is apparent in this case."[38] It is significant that X-rays of Baryulgil miners could show evidence of asbestos-related disease within the first five years of the mine's operation, when asbestosis is usually associated with prolonged exposure to very high levels of fibre. This finding is consistent with oral evidence from former miners who remember the mine as being filthy. Presumably the mine's owners, James Hardie and Wunderlich, knew of these X-ray reports for it is unlikely the Aboriginal men involved would have voluntarily attended the hospital in Grafton.

In its submissions before the inquiry, the ALS cites numerous cases of asbestos-related disease among the former miners.[39] The cases include a number of deceased subjects, whose death certificates made no mention of asbestosis but for whom there is evidence of that disease. The first case is that of a jackhammer operator, who worked at the mine from 1944 to 1966, and who died in 1969 at the age of 46.[40] His death certificate makes no mention of asbestosis, and the cause of death is given as heart failure. The man had been certified as an asbestosis sufferer by the DDB some months before his death, and he was living on a disability pension when he died in December of that year. During the last years of his life the man was severely disabled and incapable of doing anything more than light domestic work. That man was, of course, Cyril Mundine. Case 2 is that of a former mill and quarry hand, now aged 57, who suffers from shortness of breath and weight loss.[41] He has a permanent cough, and is severely disabled by a respiratory disorder. As yet, there is no evidence from X-rays of any physical abnormality other than pleural plaques. There is some controversy in medical and legal circles as to what these signify. Pleural plaques are found invariably in early cases of asbestosis, but in themselves they do not constitute proof of that or any other disease. There are a further twelve cases cited in the submission, all of whom are chronically ill from respiratory disorders.[42] Although all these men are disabled, few cases show substantive evidence of asbestosis. This, in itself, is no comfort because, in most instances, the latency period for this disease has yet to be reached. It is not uncommon for asbestosis to be invisible on X-rays until twenty years after first exposure and, even then, cases can easily be overlooked because of the difficulties of diagnosis. Of the three autopsies performed on Baryulgil miners prior to 1979, two showed evidence of asbestosis. According to the Aboriginal Medical Service (AMS) this represents a better guide to the incidence of disease than do X-rays or death certificates. The AMS claims to have traced fifty-two surviving miners. Of these, two have asbestosis and a further twelve have symptoms consistent with the early stages of the disease.[43]

In its submission, the AMS is careful to point out that the real impact of exposure to asbestos at Baryulgil mine is yet to be felt, and that the few deaths recorded to date are merely a prelude to what lies ahead. Its authors also emphasize that evidence from death certificates is of little value, unless the examining physician has access to a thorough work history of the patient. In most cases asbestosis may be a competing cause of death which is disguised or overlooked. There is much anxiety in the Baryulgil community about lung disease. That anxiety has not been allayed by the health studies carried out by the New South Wales Division of Occupational Health.

The first major health study at Baryulgil was carried out by the New South Wales Health Commission in 1977, following the death of Andrew Donnelly.[44] The report was commissioned because of concern about possible disease within the community. It contains some valuable information and some recommendations which were subsequently implemented. In this first report the authors sought to trace all miners who had worked at Baryulgil for a period of more than twelve months. The records of 208 men were uncovered and of those 105 were examined. The study relied upon a control group of 197 men from the same region in New South Wales who had not worked at the mine. From each subject information was gathered on work history, residential history and smoking habits. A complete medical check was carried out on each man, including X-rays and lung-function tests. Of the ex-miners, nearly half, 42 per cent, had worked for less than one year at the mine and a further one-third had worked for a period of between one and three years. Only 8 per cent had more than ten years' experience. The researchers discovered no major differences between the ex-miners and the control group in terms of lung function and no miner was found to have a diagnosis of asbestosis. However, an analysis of the sixty-seven identified deaths among former miners revealed that of the three cases where autopsies had been performed two showed the presence of asbestosis. It was also discovered that 11 per cent of the deaths among the group were attributable on the death certificates to respiratory disease, a non-specific diagnosis which may or may not include asbestosis.

The most significant finding in the report concerns the presence of bronchitis among the former miners and the evidence of X-ray abnormalities. The X-rays revealed abnormalities in 11 per cent of the ex-miners, but none among the control group. The abnormalities include pleural thickening, fibrosis and pleural plaques. Even more significant was the existence of chronic bronchitis among 70 per cent of the miners, which was also found in only 40 per cent of the controls. The figure for the control group is, in itself, alarming but, in the case of Baryulgil, the study reveals that nearly all subjects suffer from what can be a debilitating disorder.

The 1977 Health Commission study was based upon the concept of exposure which, in the case of Baryulgil, is impossible to quantify. Individual jobs within the mill and the quarry varied greatly in the degree of exposure. They also varied over time in the amount of fibre to which each worker was subjected. The report found that exposure for baggers probably exceeded then current standards for all periods up until 1977. Even so, the authors failed to discover any significant difference between mortality rates at Baryulgil and those found in other communities in rural New South Wales. The report recommends that the people at The Square be made aware of the dangers posed by asbestos, and that in future the disposal of waste be carefully monitored. There is also a recommendation that, in future, autopsies be carried out and that a health register be established. These findings are not surprising, given the obvious and sometimes admitted handicaps faced by the authors of the report. Information on past illness at Baryulgil is non-existent and the group of ex-miners used in the study is heavily weighted in favour of men with only a brief term of exposure. Furthermore, although the authors fail to acknowledge the fact, the latency period between the date of exposure and the survey is insufficient in most cases for the emergence of asbestosis. In various respects the 1977 study is an inconclusive document, which answered few of the questions to which it was addressed.

In 1979 a second study was carried out by the Division of Aboriginal Health in conjunction with the Divisions of Occupational Health and Health Services of New South Wales.[45] Like its predecessor, this study was intended to discover the presence, if any, of asbestos-related disease at Baryulgil. The conclusions reached in this study are much the same as those found in the previous work, with the authors failing to detect any variation in mortality rates between Baryulgil miners and the control group. There was some evidence of impairment in lung function among miners, but this could not be distinguished in its origins from the effects of cigarette smoking and the presence of chronic bronchitis. Because of the inconclusiveness of these first two studies, further research was carried out in 1981 and 1982 by the Department of Health.[46] The first involved a re-examination of the 1977 group and, in the second, forty-one ex-miners were given a thorough medical check. The 1981 study revealed once again the presence of pleural plaques in the case of seven subjects, and it found that chronic bronchitis was a widespread problem. Both the 1981 and 1982 studies demonstrate limitations inherent in carrying out research with subjects drawn from a population which suffers from poor health. Aborigines in New South Wales have such a high mortality rate and such a low life expectancy it is very unlikely that a work carried out on a segment of that population could show a marked variation in mortality or morbidity. For such an effect to be visible the Baryulgil people would have to be appallingly ill. The data on bronchitis does, in fact, show a variation but as yet the epidemiological data on other respiratory diseases is unreliable and inconclusive. That does not mean that the former miners and their families are free from illness.

James Hardie has remained adamant that there is no evidence of ill-health among the Baryulgil community resulting from employment in the mine. It is also convinced that the community itself is not at risk from environmental contamination. The company has based its opinion upon the results of the studies carried out by the New South Wales Department of Health. It also bases its judgment upon the belief that the mine was, at all times, properly managed and that workers were never exposed to high levels of asbestos fibre. The presumed absence of disease among the miners signifies, to the company, that there can be no risk whatsoever to the people at The Square as all asbestos diseases are dose-related.[47] This, of course, ignores the circumstantial evidence of disease and death in the past and the influence of the latency period on the men who worked at the mine during the 1960s and 1970s. The company's position was bolstered at the inquiry by the testimony of Dr G. Field, of Prince Henry's Hospital, Sydney, and also of the Dust Diseases Board.[48] Dr Field presented an elaborate submission to the inquiry which is, in effect, a review and critique of evidence from the Doctors Reform Society. The status of Dr Field's evidence is more interesting than its content, which is, to say the least, highly technical. In cross-examination, Dr Field acknowledged that his submission was prepared at the request of the James Hardie organization, and it is unclear whether Dr Field was paid a consultancy fee for his work, which contains much original research. Curiously, none of the inquiry members sitting raised the question of the conflict of interests between Dr Field's position with the DDB, which is mentioned on the title page of his submission, and his acting as a consultant for one of the parties to the Baryulgil controversy. Dr Field claimed that his appearance before the inquiry was in the capacity of a private citizen: he wished to be viewed as an impartial physician and not as a "James Hardie person".[49] Having thus defined the terms of his appearance, Dr Field then went on to make comments on the Doctors Reform Society submission which he deemed a political document.[50] He did not seek to justify or to explain this departure from his stated area of competence, and no member of the inquiry sought to question his expertise in carrying out such a critique. Dr Field also expressed the opinion that there is no evidence of a risk to the ex-miners and that the chronic bronchitis found at The Square is quite normal for an Aboriginal community. The question as to what is normal for an Aboriginal community is vital to the Baryulgil controversy but, unfortunately, it received scant attention during the New South Wales inquiry hearings.

The low health status of Aboriginal people in New South Wales was the subject of a report completed in September 1983.[51] Most Aborigines in the state live within a universe of low education, poor health, unemployment and poverty. Their immediate physical environment is one of inadequate sanitation, poor housing, polluted water and a lack of waste-disposal facilities. The provision of health services to such communities cannot hope to eradicate illness, high infant mortality, and high morbidity rates for adults and children. Unfortunately, there is little reliable data on the health of Aborigines. In the past, most states and the federal territories did not distinguish between Aborigines and other Australians in gathering health statistics. And, until recently, it was not possible to obtain separate data on mortality because Aboriginal identity was not cited on death certificates. A recent study of mortality in New South Wales country regions shows the life expectancy at birth for a male as 48.1 years and between 55 and 57 years for a female.[52] In both cases, in comparison with the life expectancy among white Australians, these figures represent the loss on average of twenty years of life for each individual. These figures are relevant to the New South Wales Department of Health studies of Baryulgil.

Each of the Health department studies was carried out on a small group living under considerable stress, a group for which there is no reliable comparative data. Even so, the research did manage to identify the incidence of bronchitis and some other respiratory disease as abnormally high. The studies were also based on an incomplete list of former miners, biased heavily in favour of the most recently employed who remained at the mine for a brief period. Because of the low unemployment rate at Baryulgil, one would expect the health of the people to be superior to that of other Aboriginal communities. James Hardie has made much of its benevolence in keeping the mine open but, apparently, the mine had no positive influence on the health of the people at The Square. The absence of autopsies, the absence of an index of exposure, and the absence of reliable medical histories on the surviving miners each compromise the survey's conclusions. It is known that asbestosis can develop after an exposed worker has left the hazardous environment. Therefore, the apparent absence of disease at Baryulgil is no protection against the emergence of a major health problem in the years to come. The appalling high mortality rates for Aborigines in New South Wales and at Baryulgil has almost certainly masked the effects of asbestos-related disease which have, thereby, escaped the notice of attending physicians. Consequently, the largely negative findings in the three surveys must be viewed in the light of the social and historical contexts in which they were conducted. The surveys themselves tell us more about the handicaps faced by the researchers than they do about the effects of asbestos exposure upon the miners and their families. They are no more reliable scientifically as a guide to asbestos-related disease than is the oral evidence about illnesses among the Mundine brothers and others who died after spending their work lives in the mine.

The immediate fate of the Baryulgil community was, finally, decided on a single issue; that issue is the present threat to health posed by the tailings at The Square and, in particular, the threat to the health of the children. Although James Hardie denies there has ever been such a threat, that view is not shared by the state and federal government authorities responsible for the community's welfare. Officers from the New South Wales Division of Occupational Health visited The Square in January 1981 and examined the area at the school that has been underlaid with tailings. The report from that visit,[53] which is signed by Mr L. Beattie, refers to the presence of a "major hazard" at the rear of the main school building. The area concerned was subject to heavy wear which soon exposed the remaining tailings. Beattie suggests that the only long-term solution would require the entire area to be sealed with bitumen. This report is consistent with an earlier work carried out in October 1979 by the Commonwealth Department of Health. Officers from the department found on that occasion that "the levels of airborne asbestos in Baryulgil are currently considerably in excess of the general urban community levels which have been causing international concern".[54] The source of this contamination was from the huge tailings heaps which lie less than a kilometre from The Square. Health department estimates place the asbestos fibre of these tailings as high as 40 per cent.[55]

In 1981, after more than three years of public debate between the community and various state and federal authorities, most of the people at The Square agreed to move to a new town site at Mulabugilmah, which is some six kilometres from Baryulgil along an unmade road. Not all residents moved, and a small number decided to stay, despite the danger to their health. In contrast to the housing at The Square, Mulabugilmah offered well-serviced solid brick dwellings which are quite luxurious. The quality of the new housing is not in question. However, the way in which the community was relocated gave the people little choice to consider alternatives, and it remains the cause for some concern among the Baryulgil people.

In a census taken in 1981 the population of The Square was

put at eighty-one. A Department of Aboriginal Affairs (DAA) census carried out two years later found that only twenty-eight people remained permanently at Baryulgil; eighty-seven residents had moved into the new village at Mulabugilmah. Despite the move, most of the children from Mulabugilmah continue to attend the Baryulgil school. Out of the thirty-four children at the school in 1983, nineteen were Aborigines with the rest coming from white families in the area. Three of these Aboriginal children lived at The Square, whereas the remaining sixteen travelled each day from Mulabugilmah. Since 1983, however, the number of people at Baryulgil has increased and, by June 1984, there were thirty-seven permanent residents living in nine individual households. Throughout that year there was further inward migration as people moved back to occupy abandoned houses. And yet, within the communities at Mulabugilmah and The Square, only two people are employed permanently. There are no jobs for young people and those who choose to stay at home are faced with the certainty of permanent unemployment.

Since 1977, over $3.5 million has been spent at Baryulgil in solving the problems left behind by Asbestos Mines Pty Ltd.[56] Apart from the money spent at Mulabugilmah and in sealing the tailings at Baryulgil, funds have been spent in the purchase of Collum Collum station, which was bought for the community in September 1979 by the Aboriginal Land Fund Commission. The station is situated 15 kilometres from Baryulgil, and it provides the only possibility for employment now or in the immediate future. It also gives the community an asset which is looked upon by the people with pride.

According to the DAA, the decision to move from The Square was freely taken by the community. As early as 1977 the department had formed the opinion that there was a real danger to health at Baryulgil.[57] The department had declined, however, to take any action which did not have the full support of the people and that is why it took a further three years for resettlement to be achieved. At Baryulgil and Mulabugilmah there is a division of responsibility between state and Commonwealth powers, with the state retaining control over the school and most health projects whereas the Commonwealth has responsibility for funding the village. The initial work carried out at the school to cover the exposed tailings was done by state authorities, but only after an approach had been made by the then federal minister for Aboriginal Affairs, Senator Baume. To some extent there has been co-operation between state and federal authorities but, in other respects, there has been a marked lack of communication between those authorities and the community itself. Evidence from the people of Baryulgil about the resettlement programme gives a very different picture from that presented to the Inquiry by Mr Brownbill on behalf of the DAA.[58]

The community was first told of the proposal for resettlement in 1979, and there was immediate opposition from both Aborigines and local whites, who feared that the school would be closed down. There was also concern that the local store, which services the community, would be placed in jeopardy. If the school were closed, then all children in the area, both Aboriginal and white, would be faced with a long and tiring bus journey into Grafton each day, a journey of more than two hours. For young children, such a prospect is unacceptable. A petition was signed by thirty-seven residents to protest against resettlement, and there followed a lengthy period of negotiation between Commonwealth authorities and the community. Those negotiations led, finally, to an acceptance by most people at The Square that there was no alternative. The decision was reached by the Baryulgil people in the belief that no funds whatsoever would be made available to improve conditions at The Square and, presumably, there would be no federal or state money for housing, sewage, waste disposal or electricity. The move, therefore, held the only chance of better living conditions, and it was on this basis that community agreement was given. However, since the move some public works have been carried out at The Square. It is these improvements and the status of Baryulgil as a traditional home which has, over the past three years, attracted families back to The Square. The Square is vital to the identity of the Baryulgil people, and the House of Representatives inquiry report recommendation that it be closed down came as a severe blow even to those families who had chosen to resettle at Mulabugilmah.

According to people remaining at Baryulgil, life at The Square is preferable to moving. In contrast to the suburban-style housing at the new village, which resembles nothing so much as a fragment torn from outer suburban Sydney or Canberra, there is plenty of room at Baryulgil. The general store is still used by both communities and it provides general goods, a mail service and a bank. Those who remain are not greatly concerned about the danger from the tailings, since they have lived with that problem all their lives. Because they have always been an independent community the intrusion of state authorities has come as something of a shock to many Baryulgil residents. The houses at The Square were built by the people themselves, unlike the Mulabugilmah village which, although providing good accommodation, is not home. Although there is no direct conflict between the people of Baryulgil and the new village, residents in both are now faced with a series of bewildering administrative intrusions into their daily lives.

Since the closure of the mine in 1979, the community has become a political entity. It has been the subject of television and radio exposés, the haunt of journalists seeking another sensational story, and the target for medical researchers in quest of evidence of asbestos-related disease. This recent history, overlaid with the combined activities of the DAA, the Aboriginal Development Commission, and state and federal Health departments, is difficult enough for an outsider to follow. For a community with a history of almost total neglect, recent events must seem incomprehensible. Any visitor to Mulabugilmah will hear the frequently expressed desire to return to live at The Square, which still remains home for all. Many people admit they would never have left if they had known that conditions there would improve, as they have with the provision of water and electricity. There is no employment at the new village and it is, in every sense, an artificial housing estate. It is merely a comfortable place in which to live, a fact which is of secondary importance to its residents.

The issue of resettlement should not be viewed in terms of the actions of particular government departments, which have behaved more or less efficiently in dealing with a difficult problem.

The fate of the Baryulgil community should be viewed as a consequence of the withdrawal of private industry from the site of the long-term social, medical and environmental problems which it has created, albeit by neglect. The company concerned has retreated, not just from an isolated Aboriginal community, but it has shifted all its asbestos activities either offshore to Indonesia or Malaysia or into new high-technology substitutes. These changes have been precipitated by international events and by a growing public awareness in Australia about the dangers posed by asbestos and asbestos products. Events at Baryulgil should also be seen in terms of the costs borne by that community, as a result of having an asbestos mine in its midst. Those costs include illness, social dislocaton, anxiety about the future of the community's children, and grief at being exiled from what is traditional land. These costs, and the more than $3 million in public money which has been spent so far at Baryulgil, raise an important question. That question concerns the duties and obligations of the former employer, James Hardie, towards the miners of Baryulgil, their wives and children. The company has, as yet, to address itself publicly to that question.

Since the closure of the mine in 1979, Woodsreef has carried out extensive mineral exploration in the district, and there remains a possibility that the mine may once again be opened. This possibility became more likely during 1984 with the development by Woodsreef of a so-called wet process for the mining of asbestos. This process would make the tailings heap at Baryulgil a valuable asset, as it contains a huge stockpile of partially milled ore. There would, of course, be no local demand for the mine's product, but the growing market in Southeast Asia is both close and rapidly expanding, thereby giving the company an outlet. Woodsreef's wet process would also avoid trade union opposition to the project which, in any case, is unlikely in an area where unemployment remains a chronic problem. The reopening of the mine would, however, do little to resolve the question of social justice for the people of Baryulgil.

The legal rights of the former miners and their families are discussed at some length in the report from the House of Representatives inquiry, which was released in October 1984.[59] The report includes a detailed assessment of all existing legal remedies, and a commentary on the possibility of the success of each. The major obstacle for the people of Baryulgil, in any action for personal injury, is the statute of limitations. At present, an action must be filed within six years from the date on which the injury was caused. Under New South Wales law there is no allowance for any action after that time, although courts have the discretion to grant a plaintiff a year's extension. The same provision applies also to relatives under the Compensation for Relatives Act. In either instance, the law requires that the action be taken quickly, once a diagnosis of injury has been established. In the case of asbestosis and other asbestos-related diseases, it is very difficult for an action to be brought, given the slow progression of disease from the time of exposure until the date of diagnosis. There are also further problems under the Compensation to Relatives Act, because of the common practice of traditional Aboriginal marriage at Baryulgil, which is not recognized under European law.

In cases where a plaintiff manages to bring an action, there are various obstacles, both practical and legal, which make success unlikely. Under law it is necessary to prove that a reasonable person would have been able to foresee, at the time of the injury, that harm would result from then current practice. In the United States and in Australia, the industry has been able to defend itself successfully by arguing that it did not have prior knowledge of the dangers posed by asbestos. At Baryulgil, the victim's task is made even more difficult by the ambiguous relationship between Asbestos Mines Pty Ltd, the operator of the Baryulgil mine, and the parent company, James Hardie. If the victims cannot penetrate this "corporate veil", then claims will have to be directed against a company, Asbestos Mines Pty Ltd, which today exists only on paper: it is a phantom entity without funds. Furthermore, the insurance licences taken out in the name of Asbestos Mines Pty Ltd may prove to be inoperative. Only in exceptional cases do Australian courts hold parent companies liable for the actions of subsidiaries. According to the House of Representatives report, this presents the major obstacle confronting potential litigants. When taken together with the problem of proving foreseeability and causation, it is unlikely that actions against James Hardie could succeed.

In its submissions and testimony before the inquiry, the ALS made clear its belief that, throughout the life of the mine, the company was aware of the dangers posed by asbestos. According to Mr Chris Lawrence, James Hardie engaged in ruthless exploitation of the Baryulgil miners and, knowingly, exposed them to the hazards of asbestos fibre.[60] Presumably, the situation at Baryulgil was far worse than at other James Hardie plants, as the company was abreast of the latest international scientific research on the dangers involved in the industry. In particular, Mr Lawrence claimed that the situation was known to the company's Environmental Control Committee.[61] Therefore, the problem was known at the highest levels within the organization. That knowledge, however, was not communicated to the manager, or to the workforce at Baryulgil. In evidence before the inquiry, Jerry Burke denied that he had ever been given warning by head office. Burke recalled that he first learned of the danger by reading *The New Yorker* magazine, and by following the Wittenoom case in the daily press.[62] Burke recalls having mentioned this literature to other people at Baryulgil, and of warning them that, if anything should happen to Andrew Donnelly, the fittest and one of the most heavily exposed of the miners, that they best be careful. But at no time did the company issue a warning to the miners, and no literature was distributed.[63] Burke's evidence is verified by Neil Walker, who became aware of the risks only after the death of Donnelly.[64] Like the other miners, Walker took no notice of the regular dust counts carried out by the company's officers which, in the later years of the mine, occurred regularly. To him, the dust was merely a nuisance, which had always been part of life at the mine. He remembers receiving no warning at any time, either from the manager, the Mines Inspectorate, or from other James Hardie officials. Walker's evidence and that of other miners is strongly disputed by the company.

Speaking before the inquiry on the company's behalf, Mr James Kelso acknowledged that James Hardie first became aware of the dangers of asbestos in the 1950s.[65] It took immediate action to protect its workforce against asbestosis, presumably the sole danger perceived at that time. Ventilation was the major weapon used to keep dust levels down, but it was not until the 1960s, when it became apparent that the problem was not going away, that James Hardie realized that ventilation alone was not sufficient.[66] Mr Kelso could not say when Asbestos Mines Pty Ltd first became aware of the danger of lung cancer, because in his view proof about the threat of cancer was not achieved by a specific date. The emergence of such knowledge was very complex. Medical personnel at head office would have read the literature about cancer published in 1955 establishing the connection with asbestos fibre and the company did, Mr Kelso admitted, have close contacts wth North American and British asbestos producers,[67] But Mr Kelso could not say if there were any exchange of medical information between those firms and Hardie. In response to the emergence of medical proof, Hardie began a medical surveillance scheme in 1969, but it did not take any specific steps to protect the workers at Baryulgil from cancer of the lung or from mesothelioma. An international conference on pneumoconiosis was held in Sydney in 1950, but no officer from the Hardie organization attended. Mr Kelso explained that the conference was largely addressed to the subject of coal-mining in which the firm had no interest. Unfortunately, in his evidence Mr Kelso failed to explain why it took more than twenty years for knowledge about asbestosis to reach company headquarters in Sydney, following the British legislation of 1931. A similar problem appears also to have plagued the company in perceiving medical discoveries about bronchiogenic carcinoma and mesothelioma.

In the case of Baryulgil, Hardie is adamant that the local manager must have warned the miners about the dangers of airborne fibre. And yet Mr Kelso now believes that no warning was necessary, because of the availability of the latest safety equipment and the regular dust counts and the periodic medical checks, all of which served to alert the workers to the need for care. Therefore, in the company's view, "It is inconceivable that employees were unaware of the dangers of asbestos."[68] The mine manager was on the distribution list for all information concerning safety and occupational health and he must, according to Mr

Kelso, have passed that information on to the miners. At a later hearing, however, Mr Kelso expressed a rather different opinion on this subject. When giving evidence before the inquiry on 7 May 1984, he claimed that such a warning had in fact been issued by the manager Mr Allan, and then by Mr Lloyd, and finally by Mr Jerry Burke.[69] Apparently, the company is unable to document the issuing of warnings to the miners if, in fact, such warnings were even contemplated. But, whatever the case, James Hardie has been adamant that the miners and the manager at Baryulgil were fully aware of the dangers they faced, and that the company took all measures necessary to protect the men from that hazard. On the question of protection, however, this claim is contradicted by the Hardie Papers, which shows that no such measures were taken.

The histories of the mining communities at Baryulgil and Wittenoom contain many of the same elements. These elements worked in combination to allow a particular kind of industrial practice to occur. Perhaps what happened at those towns was unique and that nowhere else in Australia were such hazardous work conditions allowed to exist. But we do know from experience in Western Europe and North America that the manufacturing and chemicals industries which have developed since 1945 have cast their own malign shadow over the lives of producers, consumers and those unlucky enough to be near waste dumps or at the scene of accidents. The tragedy at Seveso in Italy in 1976, the contamination of Love Canal in New York State, the fall-out from the pesticides industry and, especially, the widespread effects of 2,4,5-T and dieldrin are now familiar public events. In each instance, the lives of workers, families and communities have been shattered by events over which the victims had little or no control. The chemical spill at Bhopal in India in 1984 is a brutal reminder of what the costs of failure or accident can be. Even closer to home, nuclear testing carried out at Maralinga is still being debated and the effects are still being felt by the victims of political toadyism from more than thirty years ago. There were no dollars to be made in the deserts of South Australia and on Monte Bello Island, but the promise of rewards of another kind worked just as surely to excite an exotic greed. Those tests were carried out at a time when the mills at Wittenoom and Baryulgil were spilling clouds of dust and asbestos fibre into the air. Only now are the casualties from all three sites being counted. These events are also connected by the lapse in time between neglect and consequence. This gulf is the distinguishing feature of modern industrial malpractice.

There are a number of particular characteristics shared by the miners at Wittenoom and Baryulgil since the workforces at both mines display various vulnerabilities. They were poorly unionized; they entered the mining industry without prior experience with which to contrast their working conditions; they had low expectations of their jobs and of their employers; and they were largely isolated from the outside world. What expectations could an Aboriginal labourer bring to his relationship with an employer? And what expectations could an immigrant labourer from Yugoslavia be said to have on arriving at a wind-swept airport thousands of kilometres from home in the middle of a new and unknown country, and without even a rudimentary command of English? In the case of Baryulgil, the miners were Aboriginal, and the employer, the trade union, and the state government authorities treated the labour force according to its market position. That position was determined by the Aboriginal identity of the men, by their lack of marketable skills, by their small number and by their social and geographical isolation. It was a labour force without a semblance of bargaining power. The men of Baryulgil compared their situation with that of other Aboriginal workers in Kempsey, Grafton and central New South Wales towns. In comparison with so many others their lot was that of a labour aristocracy. They lived and worked on traditional land, free from outside interference. But, in truth, their choice was to work for Asbestos Mines Pty Ltd or to work for no one.

The miners at Wittenoom were mostly recent immigrants. Like the men of Baryulgil, they had little bargaining power in the labour market and little chance to make demands of their employer. Unlike the men of Baryulgil, they were mobile, and work in the mine offered the opportunity to make sufficient money to get a start elsewhere. Their relationship to the mine owners and to government instrumentalities was one of almost complete anonymity. They were of little interest to anyone, including the trade union. The men themselves were discouraged from seeking to improve conditions, because of the brevity of their employment in the mine. Whatever changes could have been wrought would not have benefited them directly, and there was no strong community among the miners binding them to those who would, in the future, take their place in the mill and at the pit face.

Both mines were unprofitable for the majority of the years of their operation. Each was owned by a major Australian company for which asbestos mining was of little importance. The mines began operating during the Second World War to guarantee a regular supply of fibre, and their usefulness was confined largely to insurance against a shortage of imported asbestos. This was more the case at Baryulgil than at Wittenoom, which was one of the few mines in the world producing commercial quantities of crocidolite. For CSR, Wittenoom was part of an adventure into the building trades industry. That diversion was never to assume great importance for the company.

The unprofitability of both mines discouraged their respective owners, CSR and James Hardie, from long-term investment in plant and dust-control equipment. In each case the mine would probably have been closed if the parent company had been forced, by the government departments responsible, to properly control the hazard of airborne dust. It is certain in the case of Baryulgil that Mines Inspectorate officers were well aware of this fact. Consequently, conditions at both mines reflected a kind of industrial anarchy. In the absence of effective outside control and influence of a concerned public, trade unions and government authorities, a system of industrial relations which rightfully belonged to the nineteenth century was able to flourish. At Baryulgil, the behaviour of Asbestos Mines Pty Ltd coincided with then current attitudes towards Aboriginal people in Australian society at large. They were marginal people without bargaining power. During the lifetime of the mine, Aborigines were not a political force anywhere in this country, which helps to explain why there is such a stark contrast between the industrial hygiene practised at that mine and the behaviour of James Hardie at its factories in Sydney and Melbourne. Even so, this does not really explain the existence of the kind of industrial practice that is suggested in the Hardie Papers.

It is possible to identify the various elements which, when combined, explain how the mines at Baryulgil and Wittenoom came to be operated in the way they were. But this aggregate of influences does not provide an explanation of why the companies concerned found such behaviour acceptable or even necessary. There is no economic motive for the absence of normal industrial hygiene, and there was no economic gain to be made from operating the mines. Perhaps benevolence was one reason for keeping the mine at Baryulgil open and, perhaps, there were men at head office who believed the company had a responsibility for the fate of the community. Unfortunately, the perception of that fate stopped short at a recognition of the known effects of asbestos fibre on the lungs of the miners. It also stopped short at a consciousness of the known environmental effects of fibre upon the health of the families of the men at both mines. On reflection, the most important fact to ponder in judging James Hardie and CSR is not the companies' behaviour during the years of mining, but their behaviour since asbestos has become a political issue in this country. It is that behaviour which is fully open to public scrutiny, and in which is displayed the kinds of motives which the ADC and the ALS have sought to expose to the public gaze.

8

The Politics of Asbestos

The degree of responsibility of any employer for the welfare of his or her employees is, under law, defined by the extent of knowledge about an existing danger. In public morality, as at law, an owner cannot be held accountable for an injury suffered by a worker which an employer could never have predicted. Therefore, in judging the behaviour of CSR and its subsidiary ABA it is necessary to identify the levels of medical knowledge about the dangers of asbestos during the life of the mine. It is also important to understand the degree of technical and medical expertise held by the company, so that blame is not assigned wrongfully.

Because of the size and economic importance of mining in this country, Australia has a strong medical literature on pneumoconiosis. By 1925 the New South Wales Coal Board had achieved a sophisticated understanding of the causes of and the means for prevention of pulmonary disease. Early breakthroughs had also been made at Broken Hill and at Kalgoorlie, and knowlege soon spread throughout those states with mining industries. In Western Australia, the Mines department was familiar with the existing literature on silicosis and miner's phthisis thirty years before the mine at Wittenoom was opened. To offset these known dangers, mining was one of the best-paid occupations in the state, and so it continued to attract a reliable workforce. Dr McNulty remembers advising one man to leave the Wittenoom mine only to be told that he was earning too much money to quit.[1] The year was 1960 and by working overtime the man was earning over £8,000 per annum. McNulty approached the management of ABA about the dust and wrote to Dr Rennie, the physician serving as a consultant to the parent firm, CSR. The Health department also made various representations but to little effect. Even without these warnings the company should have been aware of the basic medical literature on asbestos which was freely available in Australia soon after its date of publication. In a judgment handed down in a case in Western Australia in October 1979 the presiding judge was adamant that ABA knew of the dangers of asbestos fibre and in his summary made the following reference to the Merewether and Price report of 1930: "It [the report] would have formed a far graver warning to the defendants' officers than their 1944–53 knowledge of silicosis. The Merewether and Price report was available for perusal in Western Australia by September 1932."[2] Officers from the company had denied knowledge of asbestosis but they were willing to admit familiarity with silicosis which had long plagued the gold-mining industry. The judge also found that the 1948 report by Merewether which warned of the threat of lung cancer was available in Perth by 1951, again in sufficient time to demand action by ABA to protect its employees. Furthermore, CSR's decision to sell the mine at Wittenoom in 1966 suggests that it was abreast of the latest medical literature on the long-term effects of asbestos as a carcinogen.

Within two years of the 1964 New York conference on asbestos and cancer, CSR had jettisoned its interests in the asbestos industry. The mine at Wittenoom was sold back to Lang Hancock and CSR also sold its share in Wunderlich to James Hardie. It is not possible to accuse the company of seeking at that time to distance itself from evidence of past crimes. In 1964 there was no immediate prospect of the calamity which ten years later was awaiting the asbestos industry in America and Britain. And yet, in retrospect, the company's strategy in withdrawing from asbestos mining and manufacture appears eminently sensible. So do many of the strategies CSR has used since the consequences of the Wittenoom mine have become public knowledge.

In June 1978, following growing public and governmental concern about the fate of the former ABA miners, CSR established the Wittenoom Trust. According to CSR this decision showed its good faith and its willingness to assist people toward whom it was under no obligation. The Trust deed makes quite clear that there was no legal reason for CSR to establish such a fund to which it would contribute $200,000 per annum over a period of ten years. The fund is a charitable institution which dispenses voluntary assistance to former ABA employees and their families living in "unusually disadvantaged circumstances". This money is not to be seen as compensation but as charity and the Trust deed makes this distinction quite forcibly. Originally, in individual cases the amount available was not to exceed $2,000 per annum which was the same amount payable under the terms of the deed to each of the councillors serving on the Trust Board in an annual honorarium. A Trust payment of $2,000 could be made only through agreement by the councillors, and the social workers hired by the Trust had discretionary powers to grant emergency relief to a maximum of $300. Dissatisfaction among asbestos victims who approached the Trust was immediately obvious and many felt they should have the right to plead their case before the councillors.[3] This view, however, was not shared by the AWU, the trade union which had been involved at Wittenoom and which retains some interest in the fate of the ex-miners. Mr Joe Isherwood, the compensation officer for the AWU and a member of the CSR Trust, defended the Trust from such criticisms. In an interview in June 1979, Isherwood claimed that CSR had in every respect complied with its obligations as an employer and that the state government, rather than the company, was responsible for the illness suffered by the miners.[4] The government had made the regulations used at the mine and there was, he argued, no evidence that ABA had not complied with those regulations. He was also critical of people who expected the councillors to hand over funds without any consideration for the welfare of those who, in the future, would have need to call upon the Trust. In Isherwood's own words, "The Wittenoom people come to us whingeing and groaning but we have to be cruel to be kind. We have a responsibility to use the Trust money wisely."[5] Mr Isherwood declined to comment on the role of his union, the AWU, at Wittenoom which, without protest, had tolerated the existence of filthy work conditions.

There has been a convergence of interests between the union and the company which is difficult to understand given the known effects of the ABA mine upon the health of hundreds of men and their families.

From its establishment in June 1978 the CSR Trust has been besieged with appeals for help. The nature of asbestos-related diseases readily destroys a family's capacity to sustain itself, and many victims find themselves in need of public assistance. By March 1983 the Trust had dispensed $608,000 in benefits to victims.[6] This figure is rather misleading, however, as it includes the costs of salaries for social workers and administrative officers. In response to the increasing demands being made upon the Trust, in 1981 CSR agreed to increase the maximum sum payable to individual applicants from $2,000 to $5,000. Despite these increases the sums involved remain minuscule, especially when seen in light of the actual numbers of potential applicants and the amounts payable in individual cases. For a family in which the breadwinner is unable to work and requires constant nursing, a sum of $5,000 gives little respite from financial strain. In fact, the continued existence of the Trust depends upon only a small number of cases receiving assistance and only a small number of applicants approaching the Trust.

The Asbestos Diseases Society has pointed out that the Trust has a number of features which make it unattractive to those suffering from asbestosis or mesothelioma. The Trust was never intended to be responsible to the people it services and the appointment of a Trustee from the ranks of the victims was done without consultation with either the ADS or the Ex-Wittenoom Residents Association. There is no right of appeal against decisions and at every turn asbestos sufferers are reminded that they are not applying for compensation but for charity. There have also been complaints about harassment of applicants. In one particular case an officer from the Trust looked through cupboards and even under the bed to ensure that an applicant was not exaggerating her poverty.[7] Surprisingly, the ADS has no direct access to the Trust, although the society has intimate knowledge of the plight of asbestos victims and some expertise in resolving the kinds of problems from which those people suffer. The ADS has frequently complained that applicants were asked to justify the way they spend their pension benefits. This results inevitably in a violation of privacy for people who, because of unemployment, are already under the scrutiny of social welfare departments. Whatever else it dispenses, dignity is not one of the commodities in which the CSR Trust deals. The Trust has, however, allowed the company to claim that it has given benefits to ABA employees, even though it is under no obligation to do so. Above all else the Trust has provided a vehicle to promote the image of a company acting in good faith, and its very existence belies CSR's knowledge about the dangers of asbestos and the way it chose to operate the mine at Wittenoom.

Many of the victims of Wittenoom have taken their claims before the Workers Compensation Board, as was the case for the widow of Granville Rees.[8] Rees died in November 1977 after having been in receipt of a pension for asbestosis. His widow claimed that his death from cancer of the lung was the result of his employment by ABA. The company accepted that cancer was the cause of Rees' death but denied that his cancer was connected in any way with his asbestosis, for which he had already been compensated. ABA argued that the cause of lung cancer is unknown and therefore the company could not be held responsible for his death. The case attracted some publicity and it was seen at the time as something of a test for future claims. The central issues in the case were those of identifying the cause of death and establishing a causal relationship between exposure and disease.

Rees had arrived in Australia in 1955 with his wife and had worked for only a brief period for ABA as a mill-hand. After five months in the Wittenoom mill he had taken a job as a driver with Brian O'Neill, who operated a carriage contract business removing tailings from the ABA mine site. The tailings were loaded on to a truck from an open shute. There would be dust everywhere and at no time was Rees warned about the danger to his health. After each day's work Rees would return home, his clothes saturated with fibre. Rees worked for O'Neill for two years before moving back to Perth. Soon after, his health began to decline; he became easily tired and he lacked energy. In 1962 Rees first began to complain to his wife of chest pains which became progressively worse. On 6 October 1966 he was diagnosed by the Pneumoconiosis Board as suffering from asbestosis. At the time he received a little less than $3,000 for his disability which was rated by the Board at 40 per cent. During the next six years Rees was forced to work shorter and shorter hours. He had no energy and he could not walk properly or bend. He suffered from a loss of breath and he occupied his days doing light housework. In the afternoons he would lie in bed. From 1975 until his death Rees was unable to work. In July 1977 he was diagnosed as suffering from lung cancer, the disease from which he died some four months later.

According to ABA, Rees' asbestosis was quite distinct from the disease which caused his death. Therefore, the company was not liable. Rees had smoked heavily and it was probable, in the company's view, that cigarettes had been the cause of his cancer rather than his employment at the mine. This view was rejected by the Pneumoconiosis Board, which awarded Rees' widow $35,000 in compensation. The Board accepted that there was a causal connection between Rees' asbestosis and his cancer. In arriving at this decision it relied upon evidence that one-third of workers in the asbestos industry who smoke will contract cancer of the lung. The Board was satisfied that it was Rees' employment which had caused his asbestosis and that this disease had in turn precipitated his final illness. As with asbestosis, Rees' cancer was due to the inhalation of asbestos fibre. Each disease shared the same cause.

It was expected that the Rees decision would clear the way for future cases against ABA, but the company immediately lodged an appeal against the award. The company claimed that at the time of Rees' death, cancer of the lung was not a compensatable disease under the terms of the Act. The appeal was heard in the Supreme Court of Western Australia in December 1981.[9] The court addressed itself to two specific questions: Was the Pneumoconiosis Board correct in finding that death was caused by asbestosis which had in turn led to cancer? and Was the Board correct in awarding damages? Justice Brinsden found that the mechanism connecting asbestosis and cancer is not understood. He was satisfied, however, that the role played by exposure to asbestos fibre is well documented. He concluded in favour of the original decision. In his reasons for judgment, Justice Burt commented at some length upon the question of the cause of death. He noted that the ultimate cause of death in many cases is "deceptive in its apparent simplicity", and that the cause of cancer is unknown. Although he found in favour of Mrs Rees, Justice Wallace also commented on the unknown cause of cancer, but this was not sufficient ground for him to find in favour of ABA. Wallace noted that the chances of a person contracting cancer are increased appreciably for those who suffer from asbestosis and that this alone was sufficient reason to uphold the award to Mrs Rees.

The Supreme Court's decision did not lead to a deluge of litigation by asbestosis sufferers, but it does demonstrate the complex medical and legal issues of causation which lie at the heart of all such cases. Asbestosis does not cause cancer of the lung, as does cigarette smoking, but there is an enhanced possibility of cancer among asbestosis patients. Both are common where there is exposure to high levels of airborne fibre and it was this shared causation which swayed the court. The problem of causation has, however, most often worked in favour of the companies as has the prolonged period of latency between exposure and diagnosis. The very nature of asbestos-related diseases eventually forced a change to existing legislation in Western Australia, but not before many victims had struggled for years to receive the most basic forms of assistance.

Laszlo Pataki arrived in Australia as an assisted migrant in January 1961.[10] He worked at Wittenoom for three brief periods from 1961 until 1966. In 1969 he was diagnosed by the Northern Territory Department of Health as suffering from asbestosis, and was referred by that department to the Western Australian Pneumoconiosis Board. Despite a positive diagnosis, the State Government Insurance Office refused to pay him any compensation because he had been resident and employed outside the state. He was advised by the AWU and by an officer of the Western Australian Department of Public Health to go to Kalgoorlie. There he was to register as a miner and then reapply to the Pneumoconiosis Board. Pataki did this but the Kalgoorlie physicians found that he did not suffer from asbestosis and was therefore unable to apply for any pension benefit. Despite declining health Pataki returned to work as a miner, and in 1975 suffered a heart attack while working underground at Kalgoorlie. He once again applied to the Pneumoconiosis Board, this time for heart disease but the Board found that he was suffering from asbestosis. He was awarded a pension which he understood at the time was for both heart disease and asbestosis. In 1979 the pension was withdrawn. When he approached an officer from the Board Pataki was informed that heart disease was not compensatable under the Act and that his pension had been for asbestosis alone. At this time Pataki was an outpatient at a Perth hospital and he immediately went on to an invalid pension. Following further medical examinations there was still some doubt if he had asbestosis. Even so he chose to reapply to the Board for a determination of the degree of his disability. After his past dealings with the Board Pataki was confused about his health and his rights to compensation.

In 1983 Pataki joined the ADS as it was the only avenue for help. He also applied at the same time for assistance from the Wittenoom Trust but received only small grants which did not improve his situation. The Trust did however offer him a one-way ticket to Hungary, his country of birth, but as a refugee from the 1956 revolution he can never return home. The Trust has helped to subsidize his rent but the welfare officer informed him that the $30 a week he paid for his accommodation was too expensive. Many of Pataki's problems and his confusion about his rights arose from the legal and administrative frameworks which surround asbestos victims in Western Australia.

The Workers Compensation and Assistance Act of 1981 provides the legal context for asbestos-related diseases. Under the Act both lung cancer and mesothelioma are compensatable. Prior to 1981, cancer of the lung was not compensatable, which of course severely disadvantaged asbestos miners among whom this disease is so common. In the case of mesothelioma the Act allows for immediate payment of lump sums which is a necessary provision for people whose life expectancy is measured in months rather than years. The amount payable with lump-sum settlements is limited to a maximum of $58,805. This amount is adjusted in line with Consumer Price Index increases and, like all awards under the Act, it is envisaged as compensation for the loss of wages. There is no allowance made for suffering or damage to lifestyle. Such an approach disadvantages asbestos victims because diseases such as asbestosis tend to gradually reduce a worker's wage-earning capacity over a long period. By the time an application for assistance is made the person will have a lowered income from which an award is to be gauged. In the case of lump-sum payments the legal costs of gaining an award fall on the victim, thereby reducing the amount received. The most important limitation within the Act, however, is the stipulation that at the time the injury was sustained an applicant must be involved in an employer–employee relationship. This immediately excludes women and children who become ill as a result of their husband's or father's employment.

The Workers Compensation and Assistance Act of 1981 allowed for the establishment of the Workers Assistance Commission which has taken the place of the Workers' Compensation Board. The commission serves as an intermediatry between the insurance companies, the employers and the Pneumoconiosis Medical Board, which is now known as the Industrial Diseases Medical Board. The commission is the final arbiter of the level of compensation that is payable, and its role is to liaise with each of the parties and to make appointments to the arbitration panel which decides the validity of claims. There is no appeal against a panel's decision, although if an applicant is dissatisfied he or she may reapply after six months. The panel's decisions are binding on all parties. Although the 1981 Act has been welcomed universally as improving the lot of asbestosis victims, the Asbestos Diseases Society has been critical of what it views as flaws in the new system.

In Western Australia, claims for respiratory diseases are relatively few in number when compared with the worker's compensation claims in other areas of industry. In 1983 there were only 180 claims in the state as a whole. According to the ADS, the small number of applications is due to the particular characteristics of asbestos disease.[11] Apart from the latency period there are numerous problems of diagnosis faced by victims of asbestosis, mesothelioma and cancer of the lung which deter applicants. In the case of the Wittenoom mine, the high proportion of immigrant workers means that many former miners have returned to their country of origin. These men may be unaware of the nature of their illness or of their rights to claim workers' compensation in Australia. Those migrants who have remained permanently in Australia often feel intimidated by the medical and legal processes and are unable to assert their rights. Many people are unaware that they have any such rights. According to ADS records, 78 per cent of those who have approached the society are migrants and of these 44 per cent rely solely upon an invalid pension for their income. A total of 62 per cent of all ADS members are in receipt of social security benefits of some kind or other whereas only one out of five of those people receives a worker's compensation benefit.[12]

A major problem confronting applicants is the assessment procedure used in deciding the degree of disability. The judgments of the Pneumoconiosis Medical Board, or, as it is now known, the Industrial Diseases Medical Board, are based purely on a medical assessment, and the Board's work is limited to matters of medical fact. The Board gives a rating to each applicant defining the supposed level of disability. The ADS points out that the concept of impairment is a medical term, whereas the problems faced by asbestosis sufferers are not just medical but include social and economic problems arising directly from a medical condition. In theory, impairment defines the ability of a person to work. A better approach would be to assess the possibility of a person ever being able to find work. Asbestosis usually brings a partial but permanent impairment which, over time, will gradually become worse. In the case of a manual worker, a disability rating of 40 per cent means that he is highly unlikely to find a job. Such a man cannot easily retrain for white-collar work in which his disability would be unimportant. The Board's determinations ignore such distinctions.

The Act of 1981 restricts claims to the three major diseases caused by asbestos, namely, asbestosis, mesothelioma and cancer of the lung. It makes no allowance for those cancers of the gastro-intestinal tract which, in the medical literature, are associated with the asbestos industry. Such an approach excludes people who may well suffer from a disease caused by their employment but which escapes the narrow definition used by the Board. Under the terms of the Act, the greater the length of time between the accident which caused the disability and the awarding of compensation, the lower the award in real value. The compensation system also excludes provision for overtime earnings which, in the case of ex-miners, depresses the amount payable to well below the actual wages available at Wittenoom which made the mine so attractive. Where lump-sum payments are made, there are various handicaps faced by recipients. To be eligible for a lump sum an applicant must have been in receipt of worker's compensation for at least six months, and he or she is required to establish a special need for such an award. The onus rests upon the applicant to justify the payment. Invariably, insurers oppose these awards and more often than not the Board has rejected applications. The Board tends to doubt the capacity of applicants to manage their own affairs. This attitude is demeaning and it is difficult for a chronically ill man or his widow to argue against articulate and reticent Board members. In any case, the entire compensation system is cumbersome in the delivery of benefits, so that many months will elapse between the date of application and the granting of an award. This serves to further disadvantage chronically ill people, and delays of one or two years have been common.

The Western Australian government, under both Liberal and Labor regimes, has adopted a progressive stance in seeking to remove some of the obvious iniquities facing asbestos victims. That approach has, to some extent, been forced upon government by the wide publicity given to the former employees of Australian Blue Asbestos. In November 1983, the Asbestos Related Diseases Bill was introduced into the Western Australian parliament. The Bill was designed specifically to assist those victims who, because of the existing statute of limitations, were barred from pursuing common-law actions. The limitations clause meant that an action had to be instituted within six years of the cause of the action occurring. The cause of action was, in turn, defined as the date at which injury or disease had been contracted and not the date at which it had been identified or diagnosed. The new legislation was designed to remove this anomaly which excluded virtually all asbestos-disease sufferers from taking common-law action. The new Act established a distinction between prospective and retrospective claims. The dividing line between these claims was set at 1 January 1984. A prospective claim is one in which the disease arose after the set date, and ordinary statute of limitation provisions will run from the date at which the disease was diagnosed or is established as being known to the victim. Retrospective claims are those in which the disease was identified before the set date or where the limitation period of six years had expired before 1 January 1984. In all cases, where necessary, the new Act extended the limitation period for a further three years. Most important of all, the Act established an upper limit for claims at $120,000. Because the assets of Midalco, the successor of ABA are a mere $337, the Act allows for the provision of funds from state revenue if the State Government Insurance Organization or any other insurance company involved is unable to meet the claim.

The Asbestos Related Diseases Act was not intended to alter the need for a plaintiff to establish negligence. Neither did the Act in substance offer any help to the majority of the victims of ABA's mine. Only after 1 January 1959 was the common law insurance held by ABA unrestricted. Prior to that date, the insurance cover available in any single case is a mere $2,000.

The Act attracted some severe criticism especially from the Asbestos Diseases Society which was quick to point out the obstacles still faced by former miners and their relatives.[13] Under the new Act, all monies received under worker's compensation would be deducted from any settlement achieved under common law. The Act brings no relief for those who contracted a disease before 1959, and there is no loosening of the criteria which plaintiffs have to satisfy in proving the negligence of employers. The ADS has also pointed out that under the Act the SGIO and thereby the state government has agreed to pay the costs of compensation which rightfully should be borne by Colonial Sugar Refineries. The final cost to state finances could be as high as $70 million.[14] The Act was, in effect, a windfall for the company and reinforced the protection it already enjoyed through the existence of the "corporate veil" separating CSR from ABA, or Midalco, as it is now called.

Since 1981 CSR and a number of other Australian firms have fared less well in a series of common law actions. In August 1981, CSR was among thirty defendants named in an American case in which over four hundred individuals claimed damages for asbestos-related disease. All the victims were former employees of Johns-Manville to which ABA had suppled blue asbestos fibre in the period from 1948 to 1965.[15] CSR were but one of a number of suppliers named in the case, which Johns-Manville had decided to defend on the grounds that it had been unaware at the time of any dangers from its products. In Australia in April 1984, the High Court overturned a New South Wales Supreme Court ruling which had disbarred a Portuguese migrant from suing his former employer for negligence. The High Court found that the statute of limitations precluding such an action after the lapse of more than five years did not apply.[16] In Victoria in August of the same year, an out of court settlement was achieved for what was to that time a record amount for mesothelioma. The settlement was for $165,000, but because it was made out of court it had no immediate effect on subsequent cases. The plaintiff had worked as a kiln operator and he had been required, as part of his job, to wear asbestos-lined gloves. He was given no warning about any danger and during his twenty years of employment he constantly breathed in the fibre which would flake off the gloves as they began to wear out. The case did suggest, however, that employers have become aware of the difficulty in defending cases involving asbestos and asbestos-containing products. It also suggests the existence of a pool of workers who, in various ways, have come into contact with asbestos and who are now suffering from a variety of serious diseases.

The plight of asbestos victims has been made all the more difficult by the intimate relationship between government and industry at both federal and state levels and by the elevated status of the major companies involved in asbestos production. There are few Australian firms with the public profile of James Hardie and CSR, and even firms with a lesser standing, such as Woodsreef, have enjoyed a cordial and advantageous relationship with government. That relationship has done nothing to endear asbestos victims to legislators. The history of the Woodsreef-owned mine at Barraba in northern New South Wales shows just how close have been the contacts between the asbestos industry and the federal government.[17] Despite the fact that dust levels at the Barraba mine were consistently above the recommended standards, the company had no hesitation in applying to the Industries Assistance Commission (IAC) for relief in 1977 and again in the following year. Chrysotile Corporation of Australia (CCA), a subsidiary of Woodsreef, had operated successfully from the middle of 1975 until August of 1977 when it began to record heavy losses. The CBA bank was the major creditor of CCA, and in the nine-month period ending 30 June 1978 the company had lost $2.6 million.[18] The company was expected to record similar losses in the immediate future and it had therefore approached the IAC for relief. In order to help CCA, the New South Wales government had gone to the extent of waiving royalty payments from November 1977 to November 1978, and the Mines Inspectorate had also chosen to ignore CCA's noncompliance with existing dust standards.

In its submission before the IAC, CCA claimed that the mine's closure would immediately cost four hundred jobs and a further six hundred people in the district as a whole would subsequently become unemployed. The company was successful in receiving federal assistance totalling $1.4 million, but this was no sufficient to save the mine. Woodsreef had no qualms in asking Canberra for $5 million, which at a time of severe economic restraint was fanciful. By the end of 1979, CCA losses totalled more than $22 million, of which the CBA bank was owed $5 million.[19] The New South Wales government was sympathetic to Woodsreef's cause because of the employment generated by the Barraba mine in a rural area where unemployment is a chronic problem. In a similar vein, the Fraser government, after initially refusing assistance, had in dramatic fashion reversed its decision and made an emergency grant of over $1 million to head off a near-certain bankruptcy. The final IAC report on the Barraba operation issued in October 1979 makes clear the impossibility of saving the company, and it is certain that the proximity of the mine to the electorates of Doug Anthony and Ian Sinclair made Woodsreef's plight more attractive to the Fraser cabinet than would otherwise have been the case.

The failure of the Barraba mine in 1979 was not the end of Woodsreef's involvement in asbestos mining in the Northern Rivers District. In February 1984, newspaper reports suggested that the mine at Barraba was to be reopened using a new so-called wet process. The company was reported to have spent almost $1 million developing this new milling technique which would enable the mine to comply with the most stringent safety standards. The new venture was to create over one hundred jobs and to cost in excess of $12 million. However, the mine which was supposed to open in 1985 has not eventuated. There were also rumours during the same year that Woodsreef would use the wet process at Baryulgil, and supposedly a submission for assistance had gone before the federal cabinet during the Baryulgil inquiry. The real attraction of Baryulgil for future development is not the old mine, which is now filled with water, but the huge mounds of asbestos-rich tailings which have been left behind by Hardie and Woodsreef. The tailings would return an immediate yield of around 10 per cent, a far higher yield than was ever extracted from the original mine.

The buffeting which companies such as CSR and James Hardie have taken in civil actions over the past five years has served only to strengthen their resolve to fight each asbestos case as best they can. Events in Britain and the United States where corporations far larger than any Australian company have been forced into bankruptcy have made James Hardie, in particular, concerned about the consequences which may follow from the loss of even a single claim. This concern made the Baryulgil inquiry far more important than would otherwise have been the case. Even if forced by public pressure and an unfavourable parliamentary report to provide direct compensation to all the people of Baryulgil, the sum involved could easily be absorbed by a firm which spends millions of dollars each year in self-promotion.

The company was fearful, however, that such a result would have set a precedent, thereby making it necesary to grant out-of-court settlements to former employees and to those people who have become ill as a result of using Hardie products.

At the Baryulgil inquiry, James Hardie's representatives argued that in 1931 his company had no knowledge about the dangers of asbestos. The British regulations introduced in that year were, so Mr Kelso explained, directed toward the asbestos-weaving industry and had no relevance to mining or asbestos cement manufacture.[20] From the late 1940s on, the New South Wales Health Commission monitored the health of workers at Baryulgil, and there was no suggestion from that authority that there was any problem at the mine. We also know that from 1965 the Hardie company and its senior medical officer, Dr McCullagh, followed developments in Britain and America regarding asbestos and health. By his own admission, Dr McCullagh was in personal contact with the British authority, Dr Margaret Becklake, and even published original material on the subject of asbestosis in the *Medical Journal of Australia*.[21] It is more difficult, however, to establish the extent to which the firm kept abreast of current research in the period from 1944 to 1965. After that date, there are various sources which prove that Hardie was deeply concerned with the issues of occupational health and safety and that its officers routinely attended international conferences on lung disease and even contributed at such forums.

Mr N. E. Gilbert, the then Manager of Research and Development for the company attended the 1968 Australian Pneumoconiosis Conference in Sydney where he discussed at some length Hardie's achievements in the field of worker safety. According to Mr Gilbert, up until 1966 there was a problem with ventilation in many Hardie plants. In order to eradicate the problem Hardie "designed most of our own dust collection equipment in our own engineering departments".[22] As a result, large ventilation systems were installed and the problem of chronically high dust levels was overcome. The company also took various other measures in redesigning aspects of the production process and in mechanizing those functions which presented a hazard to employees. As Mr Gilbert explained, "We have decided that all our socks must be mechanically shaken, it is absolutely essential that no operator should go into the bag area and we must not depend upon an operator shaking bags at regular intervals to keep them clean."[23] The socks to which Mr Gilbert refers were a constant problem at the Baryulgil mill where no effort was made, for a further seven years, to have mechanical shakers installed. At its suburban plants, however, the company was satisfied that these measures had been successful and Mr Gilbert was confident that the kinds of dust levels which had been present at Wittenoom had never existed in Hardie plants.[24] In the mid-1960s the company had established its own industrial hygiene department under the supervision of Mr John Winters, whose job it was to monitor dust levels. Mr Winters was also involved directly at Baryulgil and he is the author of many of the Hardie Papers.

At the 1968 conference attended by Mr Gilbert, Dr McCullagh was also a participant and spoke at some length on the steps being taken to protect Hardie employees against asbestos dust. Dr McCullagh explained that the company was in the process of introducing periodic medical examinations of all employees and that almost half of the workforce had already been examined. This scheme had been welcomed by the trade unions and by the workforce and not one man had refused to participate.[25] On the basis of the statements by Mr Gilbert and Dr McCullagh the situation at Hardie plants appeared to be well in control by 1968. If accurate, this is in sharp contrast to what is known from the Hardie Papers about conditions at Baryulgil.

In 1981, the Federated Miscellaneous Workers Union prepared a detailed study of occupational hygiene at Hardie asbestos cement plants in New South Wales, Victoria and Tasmania which tends to support the optimism of Mr Winters and Dr McCullagh. There was more than a decade, however, separating the 1968 Pneumoconiosis Conference and the union study, and the company's performance at Baryulgil suggests the FMWU project may have been unduly optimistic.[26] This particular study has been cited frequently by the company as evidence of its achievement in cleaning up its plants, for the report contains barely a note of criticism. The study, which was prepared for the FMWU by Mr Noel Arnold, found that fibre levels were satisfactory in all but the Mossvale plant, although there was some problem with the feeder operation at Sunshine. Arnold was particularly impressed with the company's expertise in monitoring fibre levels, and he comments that it is the only Australian firm to undertake sophisticated research in the area.[27] He also found that respirators were available and that men engaged in dusty jobs were encouraged to wear them at all times. By 1982 the company had already decided that its future lay outside the asbestos industry and many millions of dollars had been spent on discovering substitutes in cement products. The Arnold study is therefore a portrait of a dying industry and it cannot be used to judge the kinds of hygiene practices which existed in Hardie factories over the preceding sixty years. To the company's credit, however, the study does suggest that the measures about which Mr Winters and Dr McCullagh were so optimistic had made Hardie plants safe, at least during these final years.

The strategy used by asbestos producers in Australia to protect themselves against continuing bad publicity, trade union opposition, and civil actions has followed much the same path as that employed by firms such as Turner & Newall in Britain and Johns-Manville in the United States. Almost without exception, the firms have denied that death or injury among employees was the result of exposure to asbestos fibre, and they have questioned the validity of medical evidence summoned in support of such claims. James Hardie, like Wunderlich and Johns-Manville, has denied that it kept its employees ignorant of the hazard posed by the inhalation of fibre and has argued that, on the contrary, it took immediate action once the danger was identified. The major companies have lobbied in the political arena to slow down the introduction of stringent controls over fibre levels in products and in the workplace. In Britain and America this has resulted in economically motivated decisions about worker health being reached in the political sphere to the benefit of the industry. Where cases have been brought to court, producers have always sought to slow down the judicial process to their advantage and to the disadvantage of the claimants whose financial resources are severely limited. The industry has also sought

to settle out of court, thereby avoiding bad publicity and the setting of legal precedents which could then be used by future
claimants. In Australia the industry has been under astute
management, which in the case of the Barraba mine allowed
Woodsreef to continue operating even though it continually
violated existing state hygiene legislation.

Over the past decade James Hardie has become one of the best-
publicized of Australian companies. During 1984 Hardie acted
as sponsor of numerous sporting and artistic events including
the Australian Opera, and the Fun Runs which are held in all
capital cities. The annual corporate budget for these promotions
is now in excess of $1.5 million.[28] James Hardie has explained
that its reason for becoming involved in such ventures is because
of its concern for community health and fitness. This involvement has occurred at the very time when asbestos has become a
political issue and promotions such as the "Life. Be In It." campaign seem as anomalous to the industry's critics as does the
cigarette industry's deep interest in sporting events. The company has also sought to have its perspective presented to a wider
public by the South Pacific Asbestos Association (SPAA) which,
until 31 December 1984, served as the spokesman for the industry
as a whole. The SPAA presented a thorough submission before
the 1980 Williams Inquiry in New South Wales into occupational
health and safety.[29] That submission, like others tabled by the
SPAA at public inquiries, presents the industry's rhetoric at its
most eloquent. To an extent it is an example of the industry
thinking aloud.

Mr Max Austin, the Executive Director of the SPAA and the
author of the submission, explained to the inquiry that in New
South Wales the industry employed some 25,000 people. He then
went on to attack the lack of perspective in much of the publicity
directed against asbestos in which, as he put it, powerful elements within the trade union movement had sought to force the
introduction of alternatives without regard for the costs. The use
of misleading information from overseas had done considerable
damage to local producers which could lead to the loss of jobs in
Australia. Mr Austin also explained that local producers such as
James Hardie had always been in advance of existing state safety

standards, and that such a company could not be faulted in its concern for the health of employees. He presented a table from Britain describing the number of accidents occurring annually in the workplace including the number of fatalities due specifically to asbestos. This table shows that in a single year in Britain there were 572 such deaths of which 350 were attributed to asbestos. Far from suggesting that the industry was safe Mr Austin's submission indicates that asbestos-related deaths accounted for well over half of all industry-related deaths, and that the industry is hazardous in the extreme.[30]

At a subsequent hearing of the Williams Inquiry, Mr Austin again appeared on behalf of the industry and explained something of the role played by the SPAA. According to Mr Austin, the association was not a promotional body and it was not concerned in the least with encouraging the use of asbestos products.[31] The SPAA had been formed in 1976 with the express purpose of coordinating efforts to eliminate risk from the workplace and from asbestos products. The SPAA was also concerned with achieving some balance in the public debate about asbestos which it viewed as being dominated by emotional and unreliable rhetoric. In giving oral evidence, Mr Austin explained that he had never been employed by James Hardie and that he was working for the industry as a whole. Neither he nor his employers denied that asbestos is a carcinogen, but if asbestos is to be labelled as such then all known carcinogens should be subject to the same constraints.[32] The asbestos industry should not be asked to act as the "front-line troops" in the labelling process when so many other products go unregulated. In Mr Austin's view, the industry was quite willing to warn consumers about the hazards of asbestos but it was not willing to scare people as the industry's critics would wish. Events at Wittenoom had been tragic and many people had become ill as a result, but the case of the Baryulgil mine was, so he explained, completely different and events at the two mines should not be compared.[33]

The SPAA also made a submission before the 1981 federal parliamentary inquiry into hazardous chemical wastes.[34] The submission is a defence of the Australian industry and it provides a counter to many of the criticisms levelled against Hardie and CSR over Wittenoom and Baryulgil. According to the submission, the industry ceased using blue asbestos the moment it became known that the material was dangerous, and it has always acted in a responsible manner in the use of asbestos and asbestos-based products. There are no dangers from environmental contamination, and sprayed asbestos insulation is hazardous only during the process of application. There is no danger whatsoever from this material once *in situ*. Similarly, asbestos cement water pipes represent no hazard to consumers, and claims that such piping may lead to gastro-intestinal cancer are groundless. In his oral evidence Mr Austin spoke at some length about the differences between blue asbestos and other types of fibre which were, so he explained, far less dangerous than crocidolite.

In this evidence and during subsequent appearances Mr Austin spoke about the misrepresentations which had been inflicted upon Australian producers by critics who are invariably poorly informed.[35] The SPAA denied that there had even been a conspiracy to suppress medical evidence about the dangers to health. The apparent failure of producers to react to medical evidence had been due to the erratic way in which that evidence had emerged over time and not because of any callousness or indifference on the part of producers. It was not until 1964 with the publication of Selikoff's paper on cancer that there was proof that asbestos was hazardous.[36] The Australian industry reacted immediately, and rigorous controls over airborne fibre and compulsory health checks were introduced by the larger firms. Major companies such as James Hardie have always followed a policy of self-regulation and have voluntarily sought to educate the public, consumers and the workforce about the misuse of asbestos products.[37] It is difficult, however, to reconcile these claims by Mr Austin, made in June 1981, with a James Hardie publication from five years earlier which denied the existence of any hazard from asbestos products. What had changed during the intervening period was not the medical literature, but the company's public face.

In a pamphlet titled *Setting the Record Straight,* which was probably first issued in 1976, the Hardie organization denies outright any risk to consumers of asbestos products. These products may carry warning labels but they are completely safe. Quoting New South Wales Deputy Premier Jack Ferguson, the document then goes on to reject the suggestion that asbestos cement piping could provoke gastro-intestinal cancer. The research on which Mr Ferguson based his opinion is not mentioned, although the company is content to quote him as an authority on a question which is still today a matter of medical controversy. The document concludes with the following statement, a statement which was soon to be proven false: "The measure of the company's commitment to the industry is seen by the recent acquisition of the asbestos cement operations of Wunderlich Ltd."[38] At the time of this purchase Hardie had already begun to move its asbestos operations off-shore with the establishment of plants in Malaysia and Indonesia. Within seven years, James Hardie Asbestos would have changed its name and the company would cease to be a user of asbestos fibre. It can only be hoped, for the sake of the community's health, that the other claims made in the document do not prove to be so unreliable.

In 1976 James Hardie produced what it termed "a background note", issued under the title *Asbestos and Health*.[39] The pamphlet was submitted to the Williams Inquiry and it was subsequently used as a promotional tool. The paper makes no concession to critics and presents denials of most of the more obvious and well-documented dangers associated with asbestos fibre. It refers to the believed association between blue asbestos and asbestos-related diseases. In Australia, asbestos cement piping is said to contain only 10 per cent of fibre and therefore poses no danger to consumers. The authors conceed that certain diseases such as mesothelioma are associated with inhaling *excessive* amounts of airborne fibre. This statement is, of course, misleading as the World Health Organization recognizes no level of exposure to blue asbestos at which mesothelioma cannot occur. Excessive exposure is therefore *a priori* any exposure whatsoever, which of course conveys a very different meaning from the one intended by the authors.

The document then goes on to claim that most cases of disease are the result of many years of exposure to very high levels of fibre which occurred before the risks to health were understood.

Lung cancer is acknowledged to be associated with asbestos and smoking and the authors argue that the association was not discovered until the late 1950s or early 1960s. In fact, Merewether's study of cancer was published in 1948 although the authors of the Hardie paper appear unaware of the existence of that seminal work. The note also disputes the association between mesothelioma and asbestos by emphasizing the occurrence of that disease in cases where there is no evidence of exposure. They omit to point out, however, that this is no guarantee of safety for those people employed in Australian factories where blue asbestos has been used. In summary, *Asbestos and Health* contains two features that are found in virtually all publications from the Australian industry. First, the authors make no admission of negligence by the company which is presented as being the captive of medical ignorance and fallable technology. Second, the problems of establishing a causal relationship between any carcinogen and cancer in a particular population are used as a political tool to defend the manufacturers. The occurrence of mesothelioma in people apparently not exposed to blue asbestos thereby becomes proof that blue asbestos is harmless. The known carcinogenic effects of cigarette smoking are likewise transformed into an argument that asbestos workers who smoke should give up smoking rather than that the industry should remove airborne fibre from the workplace. Finally, the difficulty of proving the existence of an environmental hazard from asbestos in suburban settings is presented as proof that consumers are safe from any such threat. Given the kind of attitudes evident in *Asbestos and Health,* it seems unlikely that any employee of James Hardie or user of Hardie products could ever take seriously the warning labels suggesting that asbestos is harmful. Until as late as 1979 the company itself was unconvinced.

It is important to remember the social context in which the industry's rhetoric has occurred. During the Williams Inquiry, evidence was presented by Dr W. A. Crawford, Director of the Division of Occupational Health in New South Wales. Dr Crawford noted that asbestos was not among those substances listed by the NH & MRC as a known carcinogen and he registered a protest against this silence. He also noted, with some horror, the ready availability of asbestos fibre in certain stores in Sydney. According to Dr Crawford, one of his officers had been sent to a Sydney hardware store following a complaint to the Consumer Protection Authority about the sale of raw asbestos. Dr Crawford explained, "When asked to be given this asbestos, the man who handled it took a shovel, dipped the shovel in a large paper bag and moved the asbestos into a smaller paper bag and dust was everywhere."[40] Presumably this was not the only store in Australia where asbestos was handled in such a way. Those endangered by exposure include not only the men and women employed directly in the industry but consumers who use asbestos in any form.

The response by James Hardie to the scandal in the United States about asbestos was decisive and has seen the company move almost entirely out of the industry. James Hardie, despite its original name, James Hardie Asbestos, was never just an asbestos company. In 1978, 95 per cent of the firm's business was located in cement products, some of which contained asbestos. By 1983 those same products accounted for less than 20 per cent of annual sales[41] and only 30 per cent of company profits. In April 1978, Hardie held in excess of three-quarters of a million tonnes of asbestos fibre in stock within Australia and New Zealand. By April 1983, the total stock had diminished to less than five thousand tonnes. This retreat has been accompanied by a rising tide of litigation from former employees and from users of Hardie products. In July 1978, in an effort to deflect public criticism about these cases, Hardie established a voluntary worker's compensation scheme whereby the firm will make payment for former employees who are totally disabled by an asbestos-related disease. The amount provided in individual cases is up to a total of $16,000, which is payable regardless of any other entitlements the worker may have received. Those eligible must, however, have been in the firm's employ for a number of years and they must have their illness certified by the Dust Diseases Board of New South Wales. The major difficulty for applicants has been acceptance by the Board, which tends to reject most claims. Unfortunately, the DDB has always refused to make public its diagnostic criteria, and applicants are unaware of the reasons

they have been rejected. Although the Hardie scheme is generous, it appears to have been motivated more by anxiety than by
benevolence.

Hardie has shed its association with the asbestos industry
through diversification into other unrelated fields of manufacture. In the principle single move, James Hardie purchased Reed
Consolidated Industries for $56.5 million, thereby gaining a
wide base in building materials, plastics and paper products.
Between 1978 and 1983 the company invested more than $200
million in new markets and new products, and also invested
heavily in a promotions campaign to create a positive public
image. In the field of cement products the company engineered
its way out of the use of asbestos. Hardie, in association with
Cape Industries, developed a cellulose substitute, and in 1981 it
purchased Australian Paper Manufacturers to guarantee a supply of that material. Through this process of diversification the
company had managed to achieve a rapid increase in profits and
a high rate of growth. Paradoxically, the end of Hardie's involvement in asbestos has been a period of prosperity.[42] The company has not, however, been able to entirely escape its past.

In May 1984, James Hardie Ltd lodged a statement of claim
with the New South Wales Supreme Court asserting that the
policies issued by QBE Insurance Ltd for public and product liability are still in force. James Hardie took this action after QBE
had given notice that it intended to rescind or to otherwise avoid
honouring contracts involving Hardie products. The QBE action
involved more than forty individual policies with various companies in the Hardie group. The insurance firm took this dramatic step because it believed Hardie had failed to disclose the
nature of the hazard associated with the group's asbestos products and therefore QBE should not be liable to honour the
policies. The outcome of the case depended upon the court's interpretation of the individual policies and the level of James
Hardie's knowledge of the dangers involved with asbestos products. The policies cover the period up to 1976 when Hardie engaged other insurers for product liability. It is significant that this
change should have occurred at a time when asbestos became a
public issue in the United States. Hardie's insurers, QBE, also

handed back to the company over $1.2 million in past premiums, thereby seeking to wash its hands of any involvement in Hardie's future. Apparently, QBE feared that it would be dragged down with Hardie as the flood of litigation from those who have used Hardie products in the past and contracted cancer or mesothelioma or asbestosis begins to rise. The American precedent is now a dark shadow for both QBE and for Hardie, as at least three major American insurance firms have gone into liquidation as a result of product liability claims. Much to the relief of producers in Australia, QBE's fears have yet to be realized, with most claims having been settled out of court.

In November 1984 Justice Rogers, the judge presiding in the QBE case, suggested the need for legislation to avoid an avalanche of asbestos claims.[43] He commented that the outcome of the QBE case would be vital for both parties and rejected the idea that in Australia there would be no repetition of the American experience which has seen the collapse of major corporations. His remarks seem appropriate if the average settlement paid by Hardie is $70,000 as reported.[44] That is why the Baryulgil story has been so important.

Throughout the Baryulgil controversy the stance adopted by Hardie has been consistent, and the firm has not wavered from its position that no harm has been done to the community. According to the firm, the work process at the mill and in the quarry was properly managed and the health of the workforce was protected by the firm's care in monitoring dust levels and introducing health surveys of all employees. Presumably, the company knew no more about the dangers of asbestos during the life of the mine than did the relevent government authorities. In any case, in the company's view, Baryulgil cannot be compared with Wittenoom: white and blue asbestos are entirely different substances. Blue asbestos is dangerous, whereas white fibre, if properly handled, presents no risk. The firm's confidence that there has been no illness among Baryulgil miners was contradicted, however, by the company's behaviour during the Baryulgil inquiry. Mr James Kelso, the Hardie representative, sought at various times during the hearings to have evidence taken *in camera* and for the suppression of documents, notably the Hardie Papers which give vital evidence about conditions at the mine.[45] The company also refused to give permission for the appearance of important witnesses such as Dr S. F. McCullagh and Mr J. Winters, who were intimately involved in monitoring health at Baryulgil.[46] Much to the disappointment of the committee chairman, Mr Jerry Hand, James Hardie refused to supply important documents on the management and running of the mine which would have been useful in giving committee members a better understanding of its operation.[47] It is difficult to reconcile this attitude with Mr Kelso's comment on the firm's commitment to an open inquiry where he said: "It is not our objective in this issue to be secretive or unhelpful."[48] This statement of principle is also contradicted by the ambiguous status of Mr Kelso's own testimony. He was a representative of James Hardie Service Industries, but the firm was not willing to be held responsible for anything he had to say.[49] His status in appearing before the inquiry was confusing to say the least.

At various points during the hearings, Mr Kelso expressed concern at the way in which the inquiry was being used by the ALS to further its own purposes and to gather information for use in future litigation.[50] He accused the ALS of holding the inquiry in contempt and of using the media to make wild allegations even while the hearings[51] were in progress. Mr Kelso also accused the ALS of deliberately seeking to mislead the committee. His main concern was that new ground could be broken by the committee which would have far-reaching effects upon the asbestos industry as a whole. In defending his employers, Mr Kelso went to some lengths to dispute the authenticity of the Hardie Papers. The parliamentary committee in its final report, however, had no hesitation in accepting the Papers as authentic although it was critical of the oral evidence given by Mr Burke and of his attitude in witholding evidence.[52] Mr Kelso was less kind about Mr Burke, whom he accused of being an unreliable witness. He pointed out that if there were dust pollution at the mine then it was the responsibility of the local manager, namely Mr Burke, to bring the matter to the attention of head office, so that the problem could be corrected immediately. This Mr Burke had not done. Furthermore, if the mine site had been cleaned up prior to inspections, it was the doing of the local manager and not of head office, which was unaware that such a practice was taking place. Mr Kelso's accusations involve important questions which can be answered only by the company itself, since only James Hardie holds the necessary documents to prove or disprove their accuracy.

The inquiry was vital to the people of Baryulgil and to the James Hardie organization, which stood to lose prestige and credibility if the final report was critical of its practice. The inquiry was also important for all those who now or in the future will be engaged in litigation against the company. In the wider political sphere, it was important for all Australian consumers because the final report defines the degree of independence possessed by the federal parliament in adjudicating issues of environmental pollution and occupational health at a time when these issues have become significant in the political arena. This explains why Hardie chose to invest such an amount of time and expertise in presenting its case at the hearings which were held through 1984. Ironically, the final report was a bitter disappointment to both the people of Baryulgil and to the James Hardie organization. The report is also disappointing for any person concerned with the questions of corporate responsibility and the capacity of the federal parliament to control the behaviour of the private sector.

The Baryulgil inquiry was established at the request of the minister for Aboriginal Affairs, Mr Clyde Holding, following representations from the ALS on behalf of the community. It was an unusual matter for consideration by a House of Representatives Standing Committee because of its specific nature and because of the gravity of the issues involved. The quality of the documents tendered at the hearings, the questioning of the witnesses, and the evidence found in the transcripts is erratic, as is the final report, which was tabled in the House on 9 October 1984. The attitude towards the Baryulgil community taken by at least one member of the committee was flagrantly unsympathetic. Mr Ian Cameron, a National Party member from Queensland, made clear his contempt for the claim that Aborigines share an unusual attachment to traditional land, and he made known his opposition to this idea during evidence from Mr Chris Lawrence on behalf of the ALS. Mr Lawrence made the observation that Aborigines are attached to land in a way that no white person could properly understand, to which Mr Cameron retorted, "That is garbage."[53] He, Mr Cameron, then went on to ridicule the idea, remarking to Mr Lawrence, "How would you know, for Christ's sake? You are as white as I am."[54] Mr Cameron's attitude during the inquiry was consistent with the views he has expressed in the House on the issues of land rights and Aboriginal health. Speaking in the House in October 1983 on an Appropriation Bill, Mr Cameron questioned the practice of granting funds to Aborigines as a solution to their problems.[55] He remarked that "such handouts should be means tested" and that Aborigines do not want to be treated as people with a separate identity. In reference to a section of the Bill dealing with the health of Aboriginal communities, Mr Cameron made the following comment: "It always amazes me that we spend $20 or $30 million a year in eradicating brucellosis in cattle, yet the dogs that live with the Aborigines are disease-ridden. I cannot see that we can ever improve the health of Aborigines unless we can replace their dogs."[56]

The chairman of the Baryulgil inquiry was Mr Jerry Hand, a member of the Socialist Left from Victoria, a man with whom one would expect Mr Cameron to have little in common. Surprisingly there were no dissenting statements in the tabled report. In fact Mr Cameron took the unusual step of praising Mr Hand as being the best chairman with whom he had worked.[57] Mr Cameron then went on to berate the minister for Aboriginal Affairs for referring the matter of Baryulgil to the committee whose time, he said, would have been better spent examining important questions such as Aboriginal education. Mr Hand's own position seems anomalous when viewed in light of his previous stance on Baryulgil. Speaking to an Appropriation Bill in October 1983, in the same week that the Baryulgil question was first referred to the committee, Mr Hand made some relevant comments. He drew attention to a number of parallels between events at Baryulgil and the possible occupational health hazards from the proposed Ranger uranium project. He expressed his strong commitment to the proper control of toxins in the workplace, and referred to the National Party's complete disregard for such issues and its desire to "help make money for their business mates". He went on to argue, "If honourable members go to Baryulgil and talk to the people there they will see the effects of the blatant disregard of people's health and welfare."⁵⁸ The report which was tabled almost exactly twelve months later hardly supports Mr Hand's comments, for it contains little criticism of James Hardie or of the way in which the Baryulgil mine was operated.⁵⁹ There is nothing in the inquiry transcripts which necessitates Mr Hand's change of heart.

The final report reads as if it were written by a number of people who did not bother to consult with each other. Consequently the three or four major subject areas that are covered in the document cannot be reconciled one with the other; nor can the final conclusions and recommendations be deduced from the body of the document. The report contains a well-researched and clearly written history of the medical literature about asbestos-related disease. This section shows that the owners of the Baryulgil mine should have been aware of the dangers of asbestos long before the mine was opened and, therefore, they were under an obligation to protect the workforce from airborne fibre. The report also covers the emergence of knowledge about the threat from cancer which again, from the first years of the mine's existence, placed the company under an obligation to protect workers. In the case of mesothelioma, the committee dates medical knowledge from the early 1960s and suggests that tailings should not have been used at The Square from that time. Once again the company is shown to have failed in its obligations toward its employees and their families. And yet at no stage in the report is a connection made between the existence of medical knowledge, the behaviour of James Hardie and the company's responsibility towards the people of Baryulgil. There are also disturbing silences, ommissions and divergences from the text of the transcripts which gives the report a curious quality. It is as if much of the most striking evidence had simply been disregarded, without explanation. For example, the oral evidence from former miners is given no credence even though the Hardie

Papers, which the committee accepted as authentic, substantiate most of the claims made by the miners and their wives. Despite this fact, the Hardie organization was displeased with the report which it believed did nothing to further either its own interests or those of the Baryulgil community.

The committee was critical of the role played by the New South Wales Mines Inspectorate and of the forewarning of inspections. However, the report exonerated the company from responsibility for the slowing down of the plant and from the clean-ups which took place before all inspections.[60] The committee also found that James Hardie played no role in providing information to the workforce on the dangers of asbestos; that no meetings were called to warn workers; that no literature was issued; that the company ignored its responsibility to educate the men.[61] Perhaps the most important question raised at the hearings concerns the dating of Hardie's knowledge about the hazard of asbestos fibre. The committee concluded that Hardie should have been aware in 1953 of the existence of the Merewether and Price report of 1930 which in graphic detail describes the dangers of asbestosis. But there is no explanation as to why the committee felt that knowledge of Merewether and Price should have taken more than twenty years to reach Australia. Apparently, the committee believed that a company involved in the asbestos industry should have no greater interest in the effects of asbestos on health than would any lay member of the public. Equally surprising is the committee's explanation as to why the company cannot be said to have exploited the Aboriginal community. Presumably because the workforce at Wittenoom was white the issue of race is not relevant to Baryulgil.[62] The equality of suffering in the two communities eliminates the question of race as an explanation for the kind of industrial practice found at the mine, since presumably Asbestos Mines Pty Ltd would have been quite prepared to treat white miners in the same way.

For all its failings, the report does present a comprehensive summary of the legal rights and avenues for action open to the people of Baryulgil. In the case of civil law claims against their employer, the miners must establish that, but for some act by Hardie, they would not have contracted the disease from which they now suffer. They must also prove that the disease was a foreseeable consequence of the employer's actions and that there were at the time practicable precautions which a reasonable employer would have taken.[63] The committee concluded that the Baryulgil miners would have little difficulty in establishing causation, foreseeability and a lack of precautions where the diseases involved were asbestosis or mesothelioma. Such action would have to be directed against the Hardie subsidiary Asbestos Mines Pty Ltd and it is unlikely that in a New South Wales court the victims would be able to bring an action directly against James Hardie, the parent company. Therefore, even though the miners may have a strong case and every possibility for success, their chances of achieving a just settlement depend finally upon the financial resources of a now defunct company. That company is unlikely to be able to satisfy a single successful claim. In the case of the dependants of deceased workers, the prevalence of *de facto* relationships at The Square would possibly debar a number of widows from pursing common law actions.

Despite the committee's confidence in the legal process to resolve the plight of the ex-miners, for the surviving relatives of the men who worked at the mine during the early years and who subsequently died there is no such avenue. The appalling ill health characteristic of all Aboriginal communities in New South Wales which, despite its unique history of full employment is shared by the people of Baryulgil, has most certainly disguised the effects of asbestos-related disease. For the relatives of those men including the Mundine brothers who are remembered at The Square as having died from respiratory disease, there is no access to a legal remedy. In the case of the surviving miners, the situation is little better because of the problem in diagnosing asbestos-related disease especially among an Aboriginal population.

Major surveys of the health of the Baryulgil community were carried out by the New South Wales Health Commission in 1977, 1982 and 1982. None of these surveys uncovered conclusive evidence of widespread asbestos disease among the miners or their families. Neither did the surveys exclude the likelihood of widespread disease in the past nor the emergence of disease in the future as the latency periods for mesothelioma, lung cancer and asbestosis expire. The committee does mention that in the three post-mortems performed on former miners all showed evidence of asbestosis.[64] The report also refers to two identified cases of mesothelioma at Baryulgil, but for reasons best known to the committee no mention is made of the significance of such a rare disease in a tiny community. Within the general population mesothelioma occurs at a rate of something like one case per million. At Baryulgil, a community of less than two hundred people, there have been two cases in the past ten years. This suggests either that the chrysotile at Baryulgil is capable of producing mesothelioma or that the residues of blue asbestos in the used bags brought to the mine site is the cause. In either instance the presence of mesothelioma raises important questions about the contamination of the local community and the failure of James Hardie to protect the community from this terrible disease. Those sections of the report dealing with these questions, or more correctly in which these questions are ignored, are idiosyncratic.

The committee claims that a full occupational profile of both mesothelioma sufferers is necessary to exclude the possibility that they may have contracted the disease elsewhere. Given the known causal relationship between mesothelioma and asbestos, this line of reasoning is difficult to understand. The committee had demanded far more stringent levels of proof from the community about the presence of a disease than would a court of law. There is no explanation anywhere in the document to explain why the onus of proof should have to rest with the people of Baryulgil and not with the Hardie organization. In the case of asbestosis the report applies much the same kind of criteria.

The circumstantial evidence against the owners of the Baryulgil mine is overhwelming both from the oral evidence and from the Hardie Papers. This evidence suggests that the mine was filthy and that no precautions were taken to warn the workers about the dangers or to protect them from the inhalation of fibre. The existing data on disease among the former miners is seriously flawed as is all such evidence about Aboriginal communities in New South Wales. At Baryulgil the actual population at risk, that is the number of men who worked in the mine, is unknown. There is little reliable data about the duration of employment, the degree of exposure or previous health, all of which make it impossible to calculate past rates of asbestos disease.

In defending itself before the Baryulgil inquiry, James Hardie made much of comparing the rates of mortality and morbidity at The Square with rates found among other Aboriginal communities in rural New South Wales. The figures used in these comparisons are unreliable and are influenced heavily by the appalling infant mortality rates which are common for Aborigines in the state. All Aborigines suffer from poor health, and the comparison involves contrasting one group of desperately ill people with another. The smallness of the Baryulgil community meant that for any difference to be visible virtually every person at The Square would have to be ill. The data does in fact show that almost all adults at Baryulgil suffer from chronic bronchitis as compared with a rate of only 40 per cent among other groups. It is important that there is no difference in mortality rates or life expectancy between Baryulgil and other Aboriginal groups, when according to the company Baryulgil was unique in offering the chance of full employment and thereby release from the grinding poverty which is endemic to other Aborigines. If the people of Baryulgil are as ill as their neighbours at Casino, Muli Muli and Collenebrai it does their former employers little credit.

The conclusions and recommendations in the Baryulgil report were poorly received by the community which had hoped for something better. The committee concluded that the way in which the mine had been operated was capable of producing asbestos-related disease but that such illness is not widespread. More correctly, the committee should have concluded that because the existing data is unreliable and fragmentary it is difficult to identify such disease with any accuracy. The absence of data is not proof that disease did not occur in the past and it is certainly no guarantee of freedom from such disease in the future. But the most damaging parts of the report for the community were those dealing with the future of The Square and the rights of the people to live where they so chose. The report argues that since the establishment of the Mulabugilma village the issue of rehabilitating The Square is no longer relevant. It also recommended that the people who had decided to remain at The Square should be encouraged to move and that " . . . every step that can be taken should be taken to discourage the continued use of The Square for residential and every social purpose and the placing of strict limitations on the residents as to the uses they can make of The Square."[65] In the view of the committee, the best outcome would be to move the people to the new village. On the date of death of the last surviving lessee the title to The Square should revert to the New South Wales government which must prohibit any further residence until satisfied that the area is safe.[66]

The people of Baryulgil were bitterly disappointed by the suggestion that The Square be abandoned. They feel that they have lost their right to live where they wish and that their attachment to The Square has been violated. Many of those who moved to Mulabugilma have expressed the wish to return to live at The Square. Many also claim they would never have left if they had known that water and electricity would be provided. They now feel betrayed and dispirited. The existing danger to health is from the asbestos tailings spread about the houses, on the roads and about the school yard during the mine's lifetime. It also comes from the airborne fibre generated by winds blowing across the huge tailings mound which lies less than one kilometre from the houses. The tailings and therefore the danger to health have been created by the companies which ran the mine. It is difficult to understand why the committee did not recommend that both James Hardie and Woodsreef should be involved in cleaning up The Square so that it be made safe for human habitation. Other firms which despoil the environment are obliged to pay the costs of restoration and yet in the case of a small and isolated Aboriginal community the same principal does not apply. There is nothing in the report to explain why the committee members felt that the people of Baryulgil should be excluded from the rights enjoyed by suburban Australians. Obviously the attitude towards Aboriginal land rights expressed at the hearings by Mr Cameron prevailed in the drafting of the final report.

The most immediate avenue open to the Baryulgil miners lies in the Workers' Compensation Act of New South Wales. The particular Act which covers their case is the 1942 Dust Diseases Act. The Act allows for compensation "to any person whose disablement for work is reasonably attributable to the inhalation of dust as a worker". The Act also allows for compensation to the dependants of any person whose death can be attributed to exposure which occurred in New South Wales. In 1967 the existing Act was amended to cover asbestosis but it was not until February 1969 that asbestosis became a scheduled disease with the New South Wales Dust Diseases Board.[67] In 1983 further amendments were made to allow for compensation for cancer of the lung and for mesothelioma as asbestos-related diseases. Claims for compensation come before the DDB, which is primarily a compensatory body that has no duties or rights of inspection. In appearing before the Board, an applicant has no right to be represented and until 1983 there was no right of appeal. The Board is sponsored by the industry rather than by individual firms, and it tends to take an industry-wide view of its role and of the rights of applicants. Where an application is made by a surviving dependant, the Board will usually award only part of the full entitlement which is then supplemented by a weekly allowance. In 1984 the maximum weekly payment under the Act was $133, but after a period of twenty-six weeks the worker receives a reduced sum calculated as a fraction of his previous wage.

The major problem confronting applicants to the DDB is the Board's definition of disability. According to the Baryulgil report, there is no evidence that the Board adopts a conservative approach in assessing individual cases, but the members did find that "it appears difficult to establish with certainty what the Board's diagnostic criteria are".[68] This leaves applicants ignorant of what will satisfy the Board. According to the Board's chairman, it always adopts the policy of giving the applicant the benefit of doubt and yet up until the end of 1984 the DDB had accepted only one case of asbestos-related disease among the twenty-eight applicants from Baryulgil. Cyril Mundine was the first such case and upon his death, which the Board accepted as being due to asbestosis, compensation was made to his widow. In the case of Andrew Donnelly, because his death certificate did not cite asbestosis as the prime cause of death, no award was made to his surviving dependants. This decision stood despite evidence from an autopsy which established the presence of asbestosis. Such practices show the limitations of the approach used by the Board in determining claims.

Dr Erich Longley, chairman of the New South Wales Dust Diseases Medical Authority, gave evidence before the Baryulgil inquiry which raised some serious questions about the regulation of asbestos in the workplace. According to Dr Longley, new asbestos regulations were drafted by the DDB committee in 1969 and these were then passed on to the relevant government authority. However, it was a further eight years before the recommendations were accepted and finally gazetted, and in that period asbestosis killed perhaps as many as one hundred and fifty employees and a further one hundred would also have contracted mesothelioma and cancer of the lung.[69] Dr Longley pointed out that in the case of vinyl chloride monomer, immediate action was taken in New South Wales, although very few cases of cancer involving that substance had been identified. He went on to comment that the major illness caused by asbestos in the future would be lung cancer, and that in the case of Baryulgil the necessary incubation period had yet to be reached.[70]

The annual reports of the DDB show a constant increase in the incidence of asbestos-related diseases. For the year ending 30 June 1981, there were 1,100 applications to the DDB, which was an increase of 400 over the previous year.[71] And yet only 78 new certificates were issued for disablement. In that same year 38 deaths were identified as being due to dust, and the report notes that new cases were expected to rise until such time as dust is fully controlled in the workplace and the effects of new measures introduced in the recent past bear fruit. In the following year there were just over 1,000 applicants, and a mere 50 new certificates.[72] The Board, however, did issue 28 certificates for mesothelioma and cancer of the lung. In that year the DDB altered its practice in issuing certificates. No longer are cases involving pleural plaques, a patching visible on X-rays which invariably is a predicator of asbestosis, accepted as certifiable, unless there is also demonstrable evidence of disability. This change was justified on the grounds of saving applicants unnecessary anxiety, and yet all it really signifies is a tightening of the diagnostic criteria being used which had already achieved a 90 per cent rejection rate for new applicants.

The difficulty of gaining acceptance by the DDB was raised at the inquiry by the ALS which argued that the Board gave applicants little chance of certification. In documents submitted to the committee the ALS cited the case of Mr Karl Schultz of Vaucluse, New South Wales, whose application has been rejected in March 1983. Sixteen months later Mr Schultz had died and a post-mortem performed at The Prince of Wales Hospital showed asbestosis and patchy interstitial fibrosis in both lungs.[73] Asbestosis had not been the immediate cause of death but the presence of the disease in an autopsy performed so soon after he had been rejected for compensation by the Board shows, at the least, the Board's fallibility in its determinations. In March 1984, the ALS submitted all relevant documents in the Schultz case to Dr Selikoff, of the Mount Sinai Medical Centre. Dr Selikoff found that, "in the United States, the decision to deny an award in the case of Schultz would be considered, at the least, idiosyncratic".[74] He went on to comment that if such an approach were to be used as a general rule it would exclude virtually all awards for occupational cancer and that from a purely scientific point of view the decision has been wrong. Mr Schultz's case may be unique but it does indicate the problems which a review body such as the DDB is confronted with in deciding the right to compensation when the available diagnostic tools are imperfect. Unfortunately, there is nothing in the Board's annual reports which acknowledges the problems resulting from these limits to knowledge, as if the Board and its members were entirely satisfied with their tools and the quality of their decisions.

The status of the DDB was also raised in the evidence given before the inquiry by two consultants to the Board. Dr Geoffrey Field gave oral evidence and also tendered a submission in which he attacked the evidence presented on behalf of the Baryulgil community. Under questionning, Dr Field admitted that his own submission had been prepared at the request of the James Hardie organization and that he had acted as a consultant to the company.[75] He wished, however, that his appearance be accepted as that of a private individual. The ambiguity of Dr Field's status as a consultant to both the DDB and the James Hardie organization was not questioned by the committee even though it involved a direct conflict of interests. Dr Field went to some lengths to proclaim his status as a private individual and not as a "James Hardie person", and yet his submission had been prepared originally at the request of the company and his formal connection with the DDB is cited on that document's title page.

Professor Brian Gandevia, who, like Dr Field, acts as a consultant for the DDB, also gave evidence before the inquiry. Professor Gandevia has for some years been a key figure in thoracic medicine in Australia and is reputed to have played an important role with the Board in the setting of diagnostic criteria and in the carrying out of tests with individual patients. Professor Gandevia has presented submissions at various public inquiries involving asbestos. In 1978 he tendered a submission on behalf of Woodsreef at the IAC inquiry into the Barraba mine.[76] In that document Professor Gandevia explains at some length his views on the Australian asbestos industry and on the threat posed to health by asbestos fibre.[77] He commented, "The asbestos industry has become deeply conscious of its responsibility to eliminate the disease hazard and much of the current research is sponsored or otherwise aided by the world asbestos industry."[78] Professor Gandevia's confidence in the motives and behaviour of the industry and in the safety of asbestos products is shared by few specialists outside the industry itself. Even so, he was willing to admit that asbestos is a carcinogen, a fact about which there is no dispute.[79] Four years later, and following growing international concern about the ferocity of asbestosis as a carcinogen, Professor Gandevia had changed his mind. He appeared as a private individual before the inquiry but unlike Dr Field he declined invitations to present a submission.[80] He explained that he was opposed to government departments taking a hard line in monitoring industry because where occupational health officers act like policemen cooperation from employers falls away and the incentive to deceive is increased. Professor Gandevia now no longer believes that asbestos is a carcinogen; he thinks it is a co-carcinogen or promoter.[81] His views are not shared by the WHO or by the IARC or even by the James Hardie organization itself, which at no time during the inquiry disputed the facility of asbestos to produce cancer or mesothelioma. The remaining controversy in the medical literature about the role of asbestos concerns its ability to produce cancer in organs other than the lung. There is no dispute about its identity as a carcinogen and Professor Gandevia's views are unusual. But they do help to explain why the DDB may at times adopt criteria of proof which, according to Dr Selikoff, would see the exclusion from worker's compensation benefits of virtually all cases of occupationally contracted cancer.

9

Safety and Profits

Since the end of the nineteenth century, there has been a shift from infectious diseases to chronic degenerative disorders as the major cause of death in Western societies. Today, the presence of these degenerative diseases is viewed as a sign of economic development, and their absence invariably identifies societies with a depressed level of economic achievement. However, eighty years of intensive medical research has done little to halt the tide of heart disease and the various cancers which, today, account for most premature deaths in Australia, Western Europe and North America. This failure has encouraged a preoccupation with the management of illness, which is now seen as holding the greatest social benefits. And yet, because such a large number of cancers are proven to be induced by environmental factors, the emphasis placed upon the management of disease seems misguided. Through the exercising of controls over those chemicals that are known to be carcinogenic, the incidence of certain cancers can be reduced. Such an approach is also encouraged by the high costs of treating cancer and by the degree of human suffering cancer can bring to its victims. And yet governmental agencies, such as the Environment Protection Agency (EPA) in the United States, have found their task of controlling the use of hazardous substances and, in particular, the regulation of the disposal of wastes frustrated by an absence of political will.

Cancer is now second on the list of the causes of death in Australia. Over the next two decades it is certain to become the dominant cause of premature death, as exposures to carcinogens in the past accelerate the incidence of cancer among a maturing population. The actual size of the problem will not be known for several decades, as the impact of the vast array of new and untested chemical products that came on to the commercial market after 1945 becomes visible. The scope of these effects spreads out far wider than the workplace through chemical pollution of the air and the casual disposal of wastes, and the transport of toxic products and their recycling. Each, in turn, touches the lives of people not employed in the chemicals industry but, who, by ill luck or geography, are drawn into its circle of influence.

There are strong historical factors which, up until now, have served to obscure the impact on health of chemical products. Cigarette smoking has been common in Western Europe for over one hundred years, but its influence upon community health did not become obvious until the 1920s. Prior to the elimination of the major contagious diseases, smoking probably had little effect on life expectancy, as most smokers would die of other causes before the effect of cigarettes became apparent. Because cancer is typically a disease of old age, longevity has been one of the reasons for the emergence of cancer as a major cause of death. A similar pattern is found in the case of air pollution, which was as much a feature of the industrial revolution in Britain 150 years ago, as of post-war Detroit or Birmingham or Melbourne today. Despite the apparent connection between air pollution, cigarette smoking and disease, it is difficult for researchers to establish a causal relationship which proves that emission of wastes into the air, or the smoking of thirty cigarettes each day, has caused a particular person or class of people to develop cancer. The diseases associated with chemical pollutants or cigarettes occur within the general population of those people who do not smoke, or who live in places far removed from the emissions of factories. There are also people who do smoke, and who live in polluted areas, who do not develop cancer. In reviewing this problem of causality, Young and his colleagues concluded, "Except in cases such as the angiosarcoma (a rare cancer) caused by vinyl chloride, where the type of cancer is rare and the association with exposure irrefutable, it is virtually impossible to differentiate a cancer caused by a specific chemical agent from a similar cancer caused by some other aetiology."[1] The toxins and carcinogens found in high concentrations in the petro-chemical industry and in metallic mining are also found in lower concentrations in the wider environment. Workers employed in tyre manufacture or in plastics processing are also exposed, as consumers and as citizens, to the same carcinogenic chemicals. One immediate problem for such people seeking compensation is to prove that their ruined health occurred in their role as employee, and not in their lives as parent or consumer or unlucky citizen.

Extensive use of epidemiological studies has managed to identify less than forty carcinogenic substances. By 1982, substances such as arsenic and certain of its compounds, asbestos, benzene, polychlorinated bi-phenols and vinyl chloride had been shown to be strongly associated with certain types of cancer in humans.[2] And yet, this return from the labours of expensive research is modest, when seen in light of the 150 substances which have been proven as carcinogenic in animals. Most substances that produce tumours in laboratory animals are cancer-causing in humans. This discrepancy suggests that many substances encountered at present in the workplace, and which have not been detected in epidemiological research are, in fact, hazardous to human health. Animal experiments indicating carcinogenesis is sufficient reason, according to the International Association for Research on Cancer, to regard such substances as constituting a risk to human health. Unfortunately, many industries using potentially dangerous chemicals employ too few people to allow for the identification of an increased risk of contracting a particular disease. To detect a twofold increase in the probability of contracting cancer among 10 per cent of a given population requires the study of 570 exposed people. But there are very few carcinogens that are lethal enough to generate a 10 per cent increase in mortality. Some carcinogens bring about an increase of only two or three cases in each 10,000 exposed workers or consumers. The research into these kinds of problems is expensive in time and funds for health research workers. It is also one area of research in which the major employers in the chemicals industry have no great interest.

In all industries, there are established limits of exposure to chemical substances that have been agreed upon as safe by em-

ployers, unions and government bodies. But these limits do not
constitute genuine thresholds at which there is no damage to
health. Rather, they represent standards at which any effect to
the health of workers has become invisible against the background of community health.

During the past thirty years, government departments in developed and Third World countries have done little to regulate the
use of toxic substances and the disposal of toxic wastes. The
inventiveness of industry has far outstripped the ability or resources of governmental agencies to monitor and control the
spread of new products. In the United States and Australia, the
use of toxins is regulated by a series of bodies which are supposed to protect the interests and health of workers and consumers.
But those bodies do not have the funds or the expertise to test
the new products for toxicity, leaving the responsibility for such
research with manufacturers themselves. Even in those cases
where a government agency may wish to act against a company,
the powers of control are too meagre, or the agencies too frail
politically, to have much success. Unfortunately, the agencies
themselves can become heavily politicized from within, in
favour of the manufacturers, as the recent EPA scandal in the
United States shows all too clearly.[3] Paradoxically, the new
chemicals, including the herbicides, are potentially dangerous to
the environment and important to the survival of human life
through the increased productivity they permit. It is this bond
that allows manufacturers to argue with some justfication that
their commercial interests coincide with the interests of the community at large, and that any limitation on their activities will
hurt the consumer. This bond is also used as an excuse by commercial interests to define what is best for the community and to
act unimpeded by public scrutiny. The sale of chemicals within
the OECD countries now exceeds $300 billion per annum. This
figure gives some indication as to the importance of manufacturers within the global economy. Australia differs from the United
States or West Germany in the fact that few new chemicals are
actually developed in Australia but are imported into the country
by subsidiaries of large international corporations, such as Dow
and Monsanto. The Australian market is small, and new prod-

ucts have invariably been tested and sold in the United States or elsewhere prior to their arrival on the local market. This gives new chemical products a legitimacy they would not otherwise have. Defenders of this system argue that if local controls were more exacting than those in the United States or Europe, then it is unlikely that new products would ever reach the Australian market.

The impact of new chemicals which have flooded the market since 1945 has been felt most immediately in the workplace. This is the first environment in which industry has been made accountable for the effects of its products upon health. But the workplace is not the only environment in which the influence of carcinogens has been present.

The first major public inquiry in Australia into occupational health and safety was the Williams Inquiry which tabled its final report in June 1981.[4] The inquiry had been commissioned almost two years earlier, and it was at the time the most comprehensive and exhaustive report of its kind. By any measure, whether in terms of the numbers of hearings, the quantity and quality of the submissions tendered, or the substance of the final report, the Williams Inquiry was a major exercise. The objective of the inquiry was to examine the ethics and practice of the work relationship in New South Wales, and to produce a method for simplifying the current system of regulation and control which would benefit both employers and employees. The inquiry revealed, however, a disturbing picture of occupational illnesses and accidents, involving a crippling burden to the public. These costs involved immediate medical expenses, individual pain and suffering, the public administration of accidents and illnesses, and the price of replacing skilled and experienced labour. The inquiry discovered that in New South Wales there was no specific legislation relating to chemicals and toxic products, and that specific regulations existed for only a small number of substances.[5] Government inspectors were found to have little authority in controlling recalcitrant employers who would take every opportunity to avoid their obligations. This administrative and legislative vacuum, when combined with the attitude of a largely apathetic workforce, meant that many employees were routinely exposed to hazardous substances.

The Williams Inquiry recommended that all workers have the right to refuse to carry out unsafe work, and that they also have the right to all necessary information about the nature of the materials they handle. The field of occupational health and safety in New South Wales was found to be covered by twenty-six separate Acts, which made the system cumbersome and largely incomprehensible to employers and labour alike.[6] The area of industrial safety had, in general, fallen within the ambit of the Department of Industrial Relations, and occupational health had come within the field of the Health Commission. Among other things, this had obstructed the collection of adequate and reliable data on occupational disease and injury. In the case of the new chemicals, many of which are carcinogenic, the absence of a coherent and informed legislative and administrative structure was potentially catastrophic. The issues involved were so serious that, in his final report, Mr Williams concluded that where an employer fails to comply with existing legislation "there seems to be no logical reason why the ultimate sanction of imprisonment should not also be available in the appropriate case".[7] Such an attitude is distant from that found in the state departments responsible for the regulation of the mines at Wittenoom and Baryulgil. It is also very distant from the attitude expressed by Professor Gandevia on how best to regulate the asbestos industry.

In its final report, the Williams Inquiry made a number of radical and far-reaching recommendations which proposed the direct involvement of labour in the monitoring and regulation of hazards in the workplace. This new ethic for the factory was to be brought about through the establishment of committees of employers and employees in all workshops with over fifty workers. The commission also called for the setting up of a single regulatory body responsible for all aspects of health and safety. Most importantly of all, the commission recommended the adoption of the ILO concept of occupational health. That concept is far broader than the traditional view of worker safety, and encompasses the mental and social well-being of employees, as well as their immediate physical protection. Critics of the Williams Inquiry pointed out that, by excluding workers employed in small factories and plants, the report thereby ignored more than half the New South Wales workforce. Little attempt was made to publicize the report at the time of its release, and neither was there much publicity in September 1982 when the Occupational Health and Safety Bill was introduced into the New South Wales parliament. That Bill saw many of the Williams Inquiry recommendations given legislative form. However, it is important to remember that the relationship between carcinogens and community health is covered, in part, only by occupational health and safety legislation. The productive process, which allows the manufacture of toxic substances, also involves the creation of huge quantities of wastes, which bring further problems for governments and public alike.

In the past, it was common for manufacturers to dump waste directly on to useless land such as swamps, marshlands and gravel pits or into disused mines. The most popular sites were those where land had little commercial value. Often such land was waterlogged and closely allied to watertables or drinking water storage. Sometimes waste would be dumped directly into deep wells and would then escape to contaminate ground water.

The reason for the haphazard dumping of toxic wastes in open sites and along highways is quite simple. The high costs of proper disposal encourages manufacturers to turn to small disposal firms, which undercut competitors conforming to federal and state standards of disposal. The costs of this price-cutting to small communities and to federal authorities can be exorbitant. This is true in both the United States and Australia. In May 1971, a waste hauler sprayed oil containing TCDD from a failed hexachlorophene plant on to the unpaved streets of a small Missouri town called Times Beach.[8] For more than a decade, the residents of Times Beach lived with TCDD concentrations as high as 100 p/b, until an investigation showed the presence of this deadly toxin. After months of haggling, it was agreed that the 2,500 residents should be bought out by the federal government at a cost of $36 million. Eight hundred homes and thirty businesses thereby passed to the ownership of Washington, which will use its freehold to make Times Beach a ghost town. The cost to public health of this lack of surveillance will not be known for twenty or thirty years, as it will take that long for any marked increase in cancer and other disorders associated with the criminal disposal of waste products to become visible in epidemiological surveys. It is possible that the $36 million spent in solving the problem at Times Beach will, in time, come to be seen as insignificant.

Australia has also experienced the major problem associated with chemicals, namely, the dangers encountered in the disposal of toxic wastes. This issue, more than any other, has awoken public attention to the hazards of the chemicals industry.

During 1981 and 1982, the Parliamentary Standing Committee on the Environment and Conservation conducted an inquiry into hazardous chemicals, and the reports from those investigations were tabled in the federal parliament in April and September of 1982. The first of those reports dealt specifically with waste, and the second with the more general problems raised by the chemical industry in Australia. At the tabling of the second document, Mr Brian Howe spoke in the House at some length on what he described as a shocking indictment of government and its lack of control over the use and disposal of toxic substances.[9] He spoke of the apathy, indifference and ineptitude on the part of government in dealing with the issue, and of the rising costs to public health and welfare which would eventually have to be met in the future. Howe emphasized the importance of overseas ownership and control of the chemicals industry, which presently sees more than 75 per cent of the manufacture in foreign hands. He also made reference to the paucity of reliable data about the volume of material produced and used. The second report simply substantiates Howe's criticisms of the lack of public control and accountability.[10]

The parliamentary committee found that of the fifty to seventy thousand chemical products released commercially in the world economy, fewer than six hundred have been evaluated for their carcinogenity. Of those substances, almost five thousand are used commonly in Australia. The exact number is not known as little reliable data on the subject is available. The picture is no less disturbing in the area of the disposal of toxic wastes.[11] The committee discovered that state governments, which under existing legislation shoulder most of the responsibility in this area, lack the expertise and funds to monitor and control the disposal of toxins. Only in New South Wales and South Australia do state instrumentalities play an active role in toxic waste disposal, whereas in the Northern Territory and in most other states responsibility falls by default upon local authorities which are ill-equipped to manage such a task. Although most Australian industry is concentrated in Sydney and Melbourne, substantial quantities of waste are generated in rural centres. This decentralization tends to further undermine the regulatory work of state instrumentalities which are, in any case, understaffed and under-financed. The committee was told of various malpractices that occur in Melbourne and Sydney, in which tankers discharged waste directly into forested areas or through sewage outlets at isolated manholes.[12] Where waste is disposed of through the means of landfill there is insufficient monitoring over standards and, in the case of Perth and Brisbane, the committee found a real danger that municipal water supplies could become contaminated. The commission also heard evidence that PCBs, a highly toxic carcinogen, had been dumped at the Brisbane City Council tip at Willawong. It is planned that, after filling, the site will be developed as a sporting facility, which raises the possibility of the exposure of children in the future to a highly dangerous substance.

The major obstacle in the control of the use and disposal of toxic substances in Australia results from the glut of Acts and ordinances. There are at least 159 such pieces of legislation which cover health and safety in the workplace.[13] The committee also discovered that in the whole of New South Wales there were only six specialist inspectors in occupational health and that in South Australia there were not sufficient inspectors to monitor factories in Adelaide, let alone in the rest of the state. In their second report the committee pointed out that, although there are forty-three substances listed by the IARC as known carcinogens, the NH & MRC lists thirteen.[14] The NH & MRC's Carcinogenic Substances Committee could not, when asked, provide the committee with a list of even those thirteen substances. The committee also found that most regulatory authorities believe that assessment of risk is a highly technical matter which should be the exclusive domain of experts. Such an attitude denies the right of those who are exposed in the workplace or through the use of products to decide if the risk is acceptable.

There is an important distinction to be made between the assessment of risk and its acceptability. Producers of toxic products refuse to allow for this distinction, and it appears from the committee's second report that the same prejudice is shared by the regulatory authorities themselves. Both would seek to have the question of acceptability rendered a technical problem. The committee rightfully points out that acceptability is, above all else, a political question which can be answered only in the public domain. It is not a question for experts. The committee findings in relation to the asbestos industry are no less critical. After studying various submissions from producers and from state authorities, it found that there had been a serious lack of regulation over the levels of exposure to asbestos fibre. The report then goes on to state:[15]

> In the 50 years since some of the major asbestos hazards were recognized, the system of self-regulation and voluntary standards, supported by advice and pressure from the State Departments of Labour, Mines and Health, has left an unacceptably high toll of asbestos-induced diseases and deaths. Industry's contribution to the reduction of health hazards under this system has been inadequate. CSR now maintains that there was not enough evidence at the time to prove a special danger beyond doubt. Whether this is so or not, there were certainly ample grounds for the gravest suspicion and there is no doubt that it was persistently told by the Health Commission that there was insufficient control of dust levels to prevent asbestosis.

There are various reasons why the chemical and asbestos industries in this country have been underregulated and why industry has been allowed to set its own codes of conduct and standards of behaviour both in the workplace and for consumers. The Australian chemicals industry is an adjunct of international companies and there are few products which come on to the Australian market without having first been accepted in other OECD states. This, of course does not excuse the lack of regulation over the disposal of wastes, nor does it excuse the lack of public accountability for the ways in which some firms have produced toxic substances with complete disregard for the ultimate effects upon employees and consumers alike. The industry has been allowed to make its own rules in part because of the attitudes that have prevailed within those federal and state departments assigned responsibility for regulating and controlling their behaviour.

Those attitudes can be seen vividly in a submission made to the Inquiry into Hazardous Substances by the Department of Science and the Environment.[16] That submission demonstrates just how close is the convergence of interests and perception between government departments, which are responsible for providing some countervailing weight to the industry, and industry itself. In mentality, at least, the submission shows that this particular department is captive of the forces it is presumed to control. The submission begins with an explanation of the importance of the chemicals industry to the national good. The industry is, so the authors claim, the cornerstone of all modern economies and, therefore, it is important that the cost of controlling chemical products be minimized.[17] The authors of the document state: "It is generally accepted that the innovation and expansion of the chemicals industry has made a major contribution to the dynamics of economic growth in industrialized and developing nations over recent decades."[18] Many of these chemicals not only protect and prolong life, but they also enhance the standard of living open to all citizens. Therefore, the issue of the regulation imposed on manufacturers can be met only at great expense. Invariably such measures require more extensive testing of products and cause delays in the bringing of new products on to the market. In the department's view, this is disadvantageous to consumers, who are denied ready access to new products, even if those products are carcinogenic or teratogenic or cause damage to the central nervous system. However, such questions as these lie outside the scope of the authors. The submission goes on to describe the smallness of the Australian market. If Australian standards were more stringent than those imposed elsewhere, this would discourage the arrival of new products and thereby, once again, disadvantage consumers.[19]

The entire submission by the Department of Science and the Environment demonstrates a placatory attitude toward industry, which is portrayed as providing benefits which would not otherwise be available. It presumes that Australian consumers have no rights to information, or to make choices about the products brought into the market place and that, in any case, they are not in a position, with regard to the global economy, to demand such rights. It is inconceivable that the department could ever be construed by the chemicals or asbestos industries as anything other than a docile friend. The same cannot be said of Australian consumers and citizens.

In the United States since 1970, a number of innovations have been adopted which were designed to protect the public against the effects of carcinogens in the environment. In March 1971, the EPA published an initial list of air pollutants. The list included asbestos fibre and, in December of the same year, the agency issued a set of standards for asbestos emission. Over the next six years there were various recommendations covering the disposal of asbestos waste and, in June 1978, the EPA issued notices setting procedures for the demolition and renovation of buildings known to contain asbestos insulation. Among other things these standards required formal notification to the agency of buildings which contain more than eighty metres of asbestos pipe insulation, or more than fifteen metres of other asbestos-based material. The notification must be made within ten days of demolition, and detailed procedures are set down by the EPA about removal and disposal. In the case of asbestos plants no visible emissions are permitted with the same provisions applying to waste disposal of fibre products. In April 1981 the EPA published new agenda for asbestos under the National Emissions Standard for Hazardous Air Pollutants, and other changes were then placed on notice for February 1982. In April 1982, further amendments were made to the Toxic Substances Act in what, by that time, had become a concerted effort to control the presence of asbestos in the general environment. In Australia there has been no comparable action by public authorities.

It is known that, since 1979, all asbestos waste generated at James Hardie's Sunshine plant was dumped at the local tip, located in Melbourne's western suburbs. Prior to that time, waste was dumped haphazardly about the factory site, and at a number of tips in the area. In either case, no special precautions were taken at Sunshine which means that some fibre escaped into the air. In Sydney there are a number of cases involving the casual dumping of waste. In March 1979, the state member for Parramatta revealed that, for some years, private home owners in his constituency had been given land fill by the local James Hardie factory for use in driveways. According to Mr Wilde, the waste contained asbestos with the result that all people living in the area had been exposed to high levels of fibre.[20] The allegation was denied immediately by the Hardie company which explained that the waste was from asbestos cement manufacture, and it was not asbestos waste as such. In South Australia asbestos waste has, for many years, been dumped routinely at the Garden Island tip in Adelaide, where no facilities were available to render the material safe. This practice continued until 1982. In each of these three cases the casual and irresponsible attitude towards the disposal of waste was ended only by public pressure brought about by vigilant journalists or members of parliament. It was not ended by the self-regulation of industry, or by the vigilance of the public authorities responsible for such matters.

The feeble attitude taken by federal and state governments towards the issue of environmental contamination is obvious in parliamentary debates. In answer to a question to the minister for Health about the potential dangers of asbestos cement sheeting in bathrooms and ceilings, Senator Don Grimes replied that there was no evidence of any danger from such material.[21] He went on to explain that the only hazard posed by asbestos was in the workplace, and that elsewhere the levels of dust were too low to be cause for concern. In the case of asbestos cement sheeting, the only threat arose during the process of manufacture and, once installed, the material is completely safe. Senator Grimes's reply is based on the probability of a person contracting asbestosis, and ignores the threat from mesothelioma or cancer of the lung from which there is no guarantee of safety. The senator's answer could well have been given by an industry spokesman, instead of by a person appointed specifically to represent the public interest, an interest which in this case does not coincide with that of the industry.

The passivity of public authorities has led to the exposure to known carcinogens of large numbers of people outside the workplace. In the case of Canberra, the problem has certain unique features because of the widespread use of asbestos as an insulating material during the 1960s and 1970s, which was a period of rapid growth in the national capital. Unlike other Australian cities, Canberra lies well above sea-level and in winter it is prone to severe frosts. This encouraged the use of home insulation, when asbestos was an obvious material for resisting cold. Although no public authority in the ACT used asbestos insulation in public housing, many people living in such houses installed their own insulation and in many cases this involved asbestos. One particular contractor named Snowy Mack worked for several years installing asbestos fluff directly into the ceiling cavities of inner suburban homes on both the north and south sides. Many homes in the suburbs of Reid, Ainslie, Deakin, Red Hill, Forrest and O'Connor are among the estimated two thousand homes serviced by this one contractor. When finally approached by officers from the Capital Territory Health Commission (CTHC) about the dangers of asbestos for both his employees and for his clients, Snowy Mack responded by immediately making his sons, who were his sole employees, partners in the business, thereby circumventing the danger of litigation. He finally ceased operations in 1979 after frequent approaches from the Capital Territory Health Commission. The material used by Snowy Mack is believed to have included blue asbestos.

It was the threat posed to the health of consumers which first encouraged Trevor Francis, the founder of the Asbestos Diseases Society, to begin lobbying for controls over the use of asbestos products. On 13 November 1978, Francis took an ABC camera crew into two Canberra homes in the suburb of Pearce to examine the insulation material in the ceiling cavities. One house was found to contain raw amosite, and the other home was insulated with pulp made from telephone books. During the subsequent television programme, Francis asked viewers to send him samples of insulation. He received between three hundred and four hundred replies. In the next two years, Francis became involved in a struggle with the Capital Territory Health Commission which took the view that asbestos insulation, if undisturbed, is completely safe. Only after much pressure from Francis did the commission agree to carry out tests for residents. These tests, however, are restricted to those contemplating renovations. Francis found that in most cases residents were unaware of the presence of asbestos and they had little idea of the possible danger to their children of mesothelioma from the inhalation of even minute quantities of fibre. The reason for this ignorance was due in large part to the trenchant attitude taken by the CTHC. The CTHC was content to reassure residents that they were safe from developing asbestosis. One person Francis spoke to at this time regularly spent his weekends in the ceiling where he kept a model train set. Others Francis met used the ceiling cavity in their homes for storage. In these and other cases the material was commonly disturbed, and yet the CTHC did nothing to warn residents of the danger.

It was largely through the efforts of Trevor Francis that the danger of asbestos insulation in private and public dwellings was raised in the federal parliament. During 1978 and 1979 there were a number of questions which eventually revealed the danger to public health. In a question on notice to the then minister for the Capital Territory, Mr Elicott, in November 1978, Mr Kerin asked what action was being taken about two Canberra public buildings which contained sprayed asbestos. Mr Elicott replied that there was no danger from that source and that he was unaware of the number of private dwellings in the ACT which may contain asbestos insulation.[22] Mr Kerin asked a similar question of the minister for Housing and Construction, Mr Groom, in March 1979, and received much the same reply. As in the first case, the minister admitted that his department was unaware of the use of asbestos materials in the ceilings of Canberra homes and, therefore, he did not know how many houses were involved.[23] In a grievance debate soon after, Mr Kerin claimed that raw amosite was to be found in the ceiling of eight thousand private dwellings and that the Capital Territory Health Commission had no power to stop further use of the material. Surprisingly, the prospect of large numbers of Canberra residents being exposed to asbestos fibre in their own homes failed to attract much public attention, and it was not until it became obvious that the problem existed also in a number of public buildings, including schools, that the issue gained political significance.

On 13 November 1979 the minister for Health, Mr Hunt, admitted that numerous public buildings, not only in the ACT but also in Sydney and Brisbane, were known to contain asbestos. These buildings included Parliament House in Canberra, the National Library, the Reserve Bank Building, Electricity House, certain buildings at James Cook University in Queensland, a number of buildings at the Canberra College of Advanced Education, Melrose High School and Watson High School, both of which are in the ACT.[24] Mr Hunt's admission showed at least that the issue was being taken seriously at the federal level, even if no effective action had been taken to warn parents of the dangers to the health of their children. The Capital Territory Health Commission had already decided that the issue did not merit action of any kind and, but for the efforts of Trevor Francis and the ADS, it is unlikely that public awareness would have improved. As early as November 1978, the commission issued a public statement in which it denied that there was any risk to health from asbestos fluff in domestic dwellings.[25] The commission declined, however, to provide any evidence to support its claim, which was at odds with the stance currently being taken by the EPA in the United States. The commission assured the public that no action was necessary and that no testing of existing insulation was called for. Documents made available from the commission under the Freedom of Information Act makes the commission's public avowals appear rather disturbing. According to one particular document written by Dr Crowe, an Assistant Medical Officer, the commission was aware that asbestos insulation was being used until as late as 1978. Dr Crowe visited one particular contractor who had been spraying asbestos for four years. This man had installed the material in hundreds of homes.[26] Dr Crowe also refers to the fact that many of the houses in the new suburbs of Kambah and Wanniassa have asbestos insulation in the roof. The tone of the document is one of urgency, and it refers to the activities of Trevor Francis as being those of a trouble-maker. A second document from March 1979, a letter by Dr MacLeod to the chairman of the commission, is even more disturbing because it establishes beyond doubt the commission's knowledge of the extent of the problem. In this letter Dr MacLeod mentions that the Capital Territory Health Commission had, since 1968, tried unsuccessfully to discourage contractors from using asbestos insulation. Dr MacLeod then went on to say, "Mr Francis mentions an estimated 2,000 to 8,000 homes contain asbestos. This could be — I say there are more than 8,000, but what does one do — examine every single home? And if we do, do we recommend everyone removes it at an unnecessary huge cost?"[27]

Dr MacLeod's reasoning was that if the problem existed and she admits that it does, then it was simply too big to deal with and should, therefore, be ignored.

It was revealed in July 1983 that several public buildings in Canberra contain significant amounts of asbestos. These buildings include The Lodge, the Canberra Theatre Centre, the Treasury, the Bureau of Mineral Resources, the Reserve Bank Building, Melrose High School, its library and assembly hall and, of course, the National Library. One of these buildings, the National Library, eventually attracted trade union action in what was to prove a successful campaign to protect employees from airborne fibre. A study carried out at the library in May 1983 by the Lidcombe Workers Health Centre showed that asbestos insulation in the ceiling could readily become airborne, and that it was deteriorating, thereby exposing those working on the upper floors to asbestos. It was also found that the fibre could enter the airconditioning system, thereby being spread throughout the entire building. In the case of Parliament House, which should have been a sensitive location, no action was taken, even though in late 1984 certain areas of the building were evacuated when excessive fibre levels were detected. As of November 1984 there was no specific legislation in the ACT dealing with asbestos, or with its removal from buildings. The CTHC has invited people to send in samples of suspected asbestos insulation to the Public Health laboratory, a small unit attached to Woden Valley Hospital, but this has not been publicized, and few housesholders have taken advantage of the offer.

In Canberra, unlike other capital cities, the trade union movement has been successful in disputes involving environmental contamination of the workplace. Throughout 1983 the Australian Clerical Officers Association waged campaigns at various public buildings, including the National Library and Industry House at Barton. In June 1984 the Taxation Department and the ACOA agreed on a number of asbestos control measures including programmes for air sampling and the evacuation of offices where fibre levels exceed 0.003 fibres/ml. The National Library, however, has seen the most important and prolonged industrial dispute over asbestos. Following repeated protests from trade unions and walk-outs by library staff, an independent inquiry was set up to report on the existing dangers from the white asbestos found in the ceiling of the fifth floor. The Builders Labourers Federation imposed bans on all extension work until the issue had been resolved. The ACOA placed full-page advertisements in *The Canberra Times* in July 1983, warning users of the library of the dangers present in the buildings and at other Canberra sites. The ACOA explained that the dispute had been brought about by government inaction, and that the union and its members had made repeated representations to the minister for the Capital Territory and to other ACT authorities and yet nothing had been done. The inquiry reported that there was a danger from the insulation both at the library and also at Watson High School. Following consultation between the Trades and Labour Council and the minister, the library dispute was ended on 9 March when the picket, which had manned the entrances for the past three months, was lifted. An advisory committee was established to resolve the question of removing or sealing the material, and to monitor fibre levels for the safety of employees and library users. The committee recommended the removal of the asbestos and the taking of daily samples, the results of which are posted for the benefit of employees. When readings are deemed excessive the building is evacuated. Unfortunately, the fire which caused extensive damage to the library at the beginning of 1985 started in the area where asbestos was being removed, and most certainly the fire would not have occurred but for that work. Throughout the library dispute the important question of why asbestos was used in the building when it was constructed long after the dangers of asbestos were known was not raised. The same question could be asked about the presence of asbestos in any number of public buildings in Australia, including the recently opened National Gallery in Canberra.

The removal of asbestos is a hazardous and expensive process involving the use of sophisticated machinery and the constant monitoring of fibre levels. The removed asbestos has also to be safely disposed of, which again is expensive. In the ACT the removal of asbestos insulation from a single building at Narrabunda College was costed at almost $700,000. According to estimates from the Department of Housing and Construction, the Narrabunda case is quite representative.[28] The removal of insulation from the Brisbane Taxation Building was estimated at $350,000, and removal work at Watson High School in Canberra was priced at nearly $600,000. The department's budget papers also list an emergency fund of $1.5 million available for urgent removal work in any one year. The minister, Mr Chris Hurford, announced in August 1984 that his department would carry out a survey of asbestos in Canberra public buildings so as to ascertain the best method for its treatment. The survey alone cost more than $1 million.[29]

The Canberra experience is not unique, and similar programmes for removal have been undertaken in Victoria and South Australia. In all cases the costs involved run into millions of dollars. In South Australia the state government has taken the bold step of introducing legislation intended to secure the removal of asbestos from all buildings. The South Australian programme is aimed specifically at removal, and not at encapsulation or sealing.[30] The final costs will involve tens of millions of dollars.

In the private sector the presence of asbestos insulation is only now becoming a political issue. From the beginning of the campaign by Trevor Francis in the ACT in 1978, it was predicted that the presence of asbestos in domestic dwellings would immediately reduce the market value of properties. Presumably, sophisticated buyers would be wary of purchasing a home containing a known carcinogen when the costs for removal would run into several thousand dollars. If CTHC estimates are correct, there are more than eight thousand such homes in the inner-Canberra area alone, and yet, buyers have paid no attention to the risk. According to the ACT chapter of the Royal Institute of Architects, the average cost of removal is around $10,000.[31] The removal process is extremely complex and the setting up of the machinery alone takes a team of workers at least four days to complete. The institute advises againt sealing in domestic dwellings, because of the need to retain ventilation through the roof. The Real Estate Institute of the ACT does not have any official position on asbestos and, according to a senior official, Mr Allan Vickers, it is rare for an agent to point out that asbestos insulation is present in a dwelling.[32] In the four years Mr Vickers had worked for the institute, asbestos has not been an issue of any importance, and it has had no effect on the market value of houses. The same experience is shared by the Real Estate Institute of Australia, and the question of asbestos has never been brought to the institute's attention, either by trade unions or by architects. This situation affords no protection for intending buyers of homes laced with a known carcinogen.

According to the registrar of the ACT division of the Australian Institute of Valuers, Mr McCann, vendors should have asbestos insulation removed before sale, in order to avoid the possibility of future litigation.[33] He would advise any vendor, when valuing a property, to find out the costs of removal, and how this may effect the final price. The institute does not have any specific policy on the issue, but it is aware of the potential problems which may arise, as consumers become more discriminating when buying properties in Canberra.

The presence of asbestos in the ceiling of domestic dwellings raises special problems for home owners. It is certain that not all people living in such homes will contract an asbestos-related disease. They are highly unlikely to contract asbestosis, and their chances of developing mesothelioma or cancer of the lung may be increased only by a factor of tenfold or twentyfold as a result of living in daily contact with airborne fibre. If there are eight thousand such dwellings and, perhaps, thirty thousand people directly involved, this may result in as few as a dozen excess cases of serious illness. The question for owners of such dwellings is whether they are willing to expose their children to an increased risk of developing an appalling disease such as mesothelioma. In making this decision, the home owners must be able to achieve an imaginative leap in projecting the consequences of a decision taken now to a date some thirty or forty years into the future, when their young children will be in middle life. This is a leap which few Canberra residents are able to make. If Trevor Francis is correct, and many homes in the ACT do contain raw amosite, then some of Canberra's children are fated to the same disease which killed Joan Joosten in 1979. Their parents, however, are unlikely to live to a sufficient age to witness the results of their passivity.

The attitude of home owners to the dangers from asbestos can be explained in a number of ways. The threat seems abstract and can be measured only in terms of a vague and distant possibility. The gap in time between indecision and consequence is extraordinarily wide. The Australian public has no tradition of militant consumerism, and they have ingrained habits of trust and naivety towards so many of the products which are used daily. This is true of pesticides and other chemicals which have come into the home over the past two decades. Governments have done little to increase awareness and, until recently, trade unions chose to ignore the same issues in the workplace. The attitude of government departments, both state and federal, has largely mirrored this passivity, and there has been no action at either level without the concerted efforts of labour or pressure groups, as was the case at the National Library. Worse still, there has been no lead given by the Australian scientific community to enliven public debate or awareness about the threat from exposure to carcinogens, including asbestos. According to Trevor Francis, the NH & MRC must itself share some of the responsibility for this situation.

In a submission presented in May 1980 before the parliamentary inquiry into hazardous substances, Mr Francis pointed out some uncomfortable truths about the behaviour of the NH & MRC over the asbestos issue.[34] Francis was highly critical of the NH & MRC's failure to cite asbestos among those substances listed as known carcinogens, even though the link between asbestos and cancer has been known for more than twenty-five years. The failure of the council to list asbestos has had important political and economic consequences, consequences of which the industry is all too aware. Among other things, it has allowed industry to refer to the NH & MRC list as proof that asbestos is not cancer-causing. Mr Francis also referred to the serious delays in the work of the council's subcommittee appointed in August 1978 to examine the asbestos problem. That committee boasted a heavy representation from the asbestos producers, which enabled industry to influence the committee's work. This, as Mr Francis pointed out, was quite improper.

The composition of the committee was also questioned in the federal parliament during a grievance debate in March 1979 by Mr John Kerin.[35] Mr Kerin argued that the presence of industry representatives on the committee was ludicrous in the light of the industry's willingness to expose Australian workers to asbestos fibre in the workplace. Why then, he asked, should the industry have the right to advise government on an issue of public health? The members of the industry, who were present on the committee, came from James Hardie and Woodsreef.

The publications from the NH & MRC on the subject of asbestos and health are not impressive documents. They tend to present a cautious stance on the causal link between exposure to fibre and cancer, with much emphasis being laid upon variations in risk between the three major types of fibre. The concept of caution preferred by the council is one in which the absence of demonstrable risk, as in the case of environmental exposure, is taken as evidence that there is no risk. That is, the council follows the exact line of reasoning favoured by industry.[36] No action is seen as being preferable to preventive action which, in the council's view, is necessary only when there are bodies to count. This concept of caution has strong commercial and political overtones. In the hands of the NH & MRC, the issue of safeguarding public health is not treated as a key element which must be balanced against the twin notions of rigorous science and the public good.

A second document, released by the council in June 1981 displays the timidity found in most of the council's public statements on the asbestos issue.[37] In its own defence, the council would no doubt argue that it is not a political body, and it would be quite improper for the NH & MRC or its members to become involved directly in the controversy. Unfortunately, there is nothing other than political knowledge for all those engaged in the asbestos scandal and that includes the council just as surely as it does industry or government. The council's reports are therefore political documents which have been used by industry to its own advantage. The NH & MRC's report of June 1981 makes a number of recommendations which are important in explaining why there has been such little public understanding of the dangers of asbestos, and why that understanding has come so late and so erratically. The report suggests six alternative forms for the labelling of asbestos products. These labels are designed to warn consumers about the hazards they faced in purchasing or using such products. Only one of these six labels mentions the word cancer and even then the reference is not prominent.[38] Once again the concept of caution employed by the council is identical with that favoured by producers. It is impossible to reconcile either with a broader understanding of the public interest. At best, the NH & MRC can be accused of failing to give leadership on what is an issue of some importance, an issue about which the council possesses some expertise.

The attitude of the NH & MRC is not shared by either the trade union movement or the National Consultative Committee on Occupational Health and Safety. After having ignored the issues of occupational health in favour of wage and condition claims, the Australian trade union movement has recently become active in the area. The ACTU has monitored the asbestos controversy and it has been successful in gaining a number of agreements with employers over the proper protection of workers from fibre. Notable accords have been reached at the Williamstown Dockyard in Melbourne regarding the conditions for removal of asbestos from naval vessels. The ACTU/VHTC Occupational Health and Safety Unit has also published guidelines for working with asbestos. The guidelines, which became available in October 1984, carry the estimate that each year in Australia more than six hundred workers die from asbestos-related diseases.[39] There is, of course, no way to substantiate such a claim and it appears fanciful in light of figures released by the Dust Diseases Board of New South Wales or by industry, which show that the number of cases is quite small. But, at least, the ACTU publication demonstrates that the trade union movement is making an effort to educate the workforce. In the absence of any action from industry or from government departments, the role of the trade unions becomes all the more important.

During 1984 the National Consultative Committee on Health and Safety released two important documents on asbestos in the workplace: *Asbestos Management* and *Controlling Asbestos Hazards*.[40] The second of these reports refers to the extent of the existing hazard from asbestos as being "enormous",[41] and goes on to call for a national strategy to protect people exposed in a variety of environments. The report also mentions the need for warning labels, which presumably would carry greater weight than the feeble suggestions put forward by the NH & MRC three years earlier. The reader of either report is left in no doubt as to the hazardous nature of asbestos and of the extensive precautions which should be taken by any person coming into contact with asbestos or asbestos-containing products. These are conclusions which producers would dispute, as would the NH & MRC. It is certain, however, that the former miners from Wittenoom and Baryulgil and their families would find some comfort, no matter how belated, from the knowledge that, at last, the threat of asbestos to health is being treated with the seriousness it deserves.

10

Conclusion

In the United States an estimated quarter of a million people will die of asbestos-related disease by the turn of the century. Those deaths will occur because, in the period from 1940 until 1979, more than twenty-seven million Americans were exposed to asbestos fibre. These estimates do not come from the radical environmental lobby, but are the result of epidemiological research carried out at the Mount Sinai Hospital in New York. They give some indication as to the importance of the asbestos scandal and the havoc which the use of asbestos has wrought upon community health. These estimates also suggest that in financial terms the eventual cost will be extraordinary. In evidence given in February 1982 before a Congressional inquiry into asbestos, an insurance official told a House of Representatives Committee that product liability suits against producers were likely to exceed $38 billion, and that they could go as high as $90 billion.[1] More disturbing than even these estimates is the prospect that the asbestos story will be repeated in other industries which have manufactured carcinogenic products.

In the United States as in Australia, trade unions have until recently shown little or no interest in occupational health and safety. This neglect was due to a number of factors, including lack of funds for independent research, lack of expertise in what are invariably highly technical debates, and the lack of a traditional role for trade unions in this area. The cost to the community of the use of asbestos has already been felt through the added burden of higher health insurance premiums, and the strain placed on hospitals and social welfare in the provision of care for the chronically ill. In July 1984, a Congressional House Labour Standards Subcommittee estimated that each year in the United States more than $3 billion is paid out in welfare, social security and Veterans' Administration benefits, which should rightfully be provided through workers' compensation. The effect is that the private sector has escaped the costs of industrial malpractice, which is then borne by the public at large.

The asbestos scandal is unlike any other public health issue of the past forty years. In scale it dwarfs individual accidents such as those at Minamata Bay in Japan or Seveso in Italy or at Love Canal in New York State. In each of those cases, the number of dead may be counted in the tens or hundreds, and none has produced thousands of casualties. Because asbestos is so common in urban environments, and because the health of so many people has been put at risk, it is necessary to make an imaginative leap in contemplating the financial costs or the political consequences associated with the use of this one mineral. The asbestos story is also unusual because of the numerous academic boundaries over which it wanders. No one discipline is capable of embracing all the territory involved, or of claiming expertise in the production of research about asbestos disease, the transmission of that knowledge, and the weighing of moral responsibility which must inevitably be made. This fragmentation has slowed down serious public debate about asbestos, and it has hampered the resolution of the issue in America and in Australia. More than anything else, it has played into the hands of industry by turning the issue into a technical debate among experts.

The industry has sought to depoliticize the controversy, as if asbestos-related disease were merely a series of random events thrown up by the past which tell us nothing about the present or the future. The scale of the use of asbestos and the vast numbers of potential victims have given producers every opportunity to blind the public as to the likely or known consequences of the use of asbestos products.

There are three dominant interpretations that are used to explain the asbestos story. From the hard left comes the idea that such tragedies arise because of a conspiracy within the industry, which was systematic and well organized. That conspiracy, supposedly, saw corporations suppress medical evidence proving asbestos was hazardous, so that high profits could be dragged from the suffering of workers and consumers alike. Capitalism is seen as evil, and the sole motive or explanation for events such as those at Wittenoom and Baryulgil is the greed of the ruling class. For each worker whose lungs were ruined by asbestos, there was a bag of dollars which some corporate giant was able to stuff into its pockets. By neglecting to introduce adequate protection for its workforce, companies such as ABA were able to reap higher and higher rewards. Unfortunately for the hard left there is no evidence that either CSR or James Hardie ever succeeded in making money from their asbestos mines and, if any criticism is to be made of those firms, it is that they were incompetent rather than malevolent capitalists.

Industry itself explains events such as those at Baryulgil as being due to the fallibility of medical and scientific knowledge. Ignorance forced upon producers by medical and other professionals is the sole reason for the asbestos scandal. According to industry spokesmen, the identification of illness and hazard is achieved through specific technical advances. Solutions to existing hazards are arrived at as new techniques and inventions, such as the Midget Impinger and its successor, the Electron Microscope, become available. The moral imperative for dust control or ventilation or the provision of respirators must come from the laboratory, and not arise from the sensibility of the board room. Industry, so we are told, can do only what scientists advise, and the acceptance or rejection of that advice is determined solely by the weight of scientific evidence, which is ethically neutral and unambiguous.

There are a number of consequences of the mode of explanation favoured by industry. Once a controversy is metamorphosed, or translated into a technical problem, the opinions and evidence of victims becomes worthless. Men and women who have worked in asbestos plants or who have consumed asbestos products are held to be incompetent as a source of reliable and objective information. They cannot be relied upon to give evidence about dust levels, or their own illness and its cause. In this context, access to information is the key to successful litigation, and this information is monopolized by industry. Invariably, there is no countervailing weight of opinion from physicians to give support to the victims' claims. That is why so few asbestos cases succeed in court.

Besides the modes of explanation promoted by industry and by the hard left, there is a third and more sophisticated interpretation of the asbestos story. In this third alternative, such tragedies are seen to arise out of a complex web of relations between capital, government, public, consumers and the workforce. Each of these sets of relations is mediated through the opinions, prejudices and actions of technicians. Those technicians are centred in state and federal authorities, in industry and in medicine at large which, in this instance, is best described as a "disabling profession". Perhaps the most important single element in the story concerns the way in which medical knowledge is produced, the conditions of its production, and the social, ideological and political circumstances of its dissemination. Without an understanding of that process, the asbestos story remains mysterious.

During the first decades of the present century, public health programmes designed to combat the spread of infectious diseases were introduced without any of the political resistance that has accompanied the struggle for controls over the use and dissemination of toxic substances. In the battles to control cholera, tuberculosis and, more recently, infantile paralysis, policymakers merely informed publics that there was a certain risk, and that particular measures were necessary to eliminate that risk. Legislators needed little encouragement to enforce such measures and there was no opposition. There were no vested interests which stood to lose money or prestige or power from the elimination of those diseases. The opposite has been true of every case involving government efforts to control carcinogens in the market place. The discovery of an association between a disease and a chemical product is a political act. Its creator and the knowledge itself are immediately embroiled in an environment where the virtues of honesty and objectivity are largely irrelevant. The mere design of the protocol for the study of an illness becomes controversial, and serves to polarize contesting parties on either side of what are simultaneously scientific and political debates. The politicization of science is used by industry to immobilize debate, by excluding the victims and the public from participation in the conflict. That exclusion is a political strategy which is rarely, if ever, justified by the nature of the explanations involved, or by the methodology physicians use in researching diseases and their cause. What needs to be done is to transfer the asbestos controversy into the public arena, so that the issue is no longer seen or defined as a technical problem for experts in the legal, medical, technocratic and bureaucratic spheres to debate unhindered by an informed public. In the United States and in Britain, this process has already begun through pressure from litigation, and by the work of consumer and public interest groups. It has not been initiated by parliaments or by public awareness of the consequences of environmental contamination by asbestos. Neither has the process been precipitated by a concern for the victims, who remain a scattered and politically insignificant group.

The asbestos industry in Australia has always been small and the total number of employees can be measured in the hundreds, rather than the tens of thousands who worked in the industry in the United States. Both the mines at Wittenoom and at Baryulgil were of little importance to the parent companies, CSR and James Hardie, and neither was central to their corporate strategies. Hardie may have, as it claims, retained the Baryulgil mine largely out of concern for the welfare of the community, or it may have been motivated by hopes of a shift in the local market that would have favoured large-scale development of the deposit. Even more likely is the possibility that Hardie kept the mine operating out of habit. Similarly, the company's failure to inform the workforce about the hazards of asbestos fibre was probably due to lethargy. The mine employed very few men; there was no effective trade union presence; the Aboriginal workforce had a low status in the community at large, a status which was reflected faithfully in its treatment by the company. From as early as 1944, the first year of the mine's operation, existing medical literature demanded caution and the introduction of various well-established safety measures and yet the company chose to ignore that necessity. Although the mine may have been of little significance during its lifetime, the Baryulgil mine and the people of The Square have since assumed a considerable political importance to the company. That community now presents a spectre of James Hardie's precarious future.

Baryulgil is a community of less than two hundred people and, at its peak, the mine employed a mere forty men. Neither the township nor the community is of any economic importance, least of all to James Hardie. But events at Baryulgil have cast a shadow across the future of the company, for they are a reminder of the forces which, over the past five years, have destroyed several American asbestos conglomerates. Baryulgil is also important because of the amount of public money that has been used in solving a problem created by a private firm. Already more than $3 million has been spent in relocating the community and in monitoring the health of the former miners and their families. Baryulgil has been a political issue in New South Wales since 1977. It has been the source of several civil claims in court, and to James Hardie it presents what may prove to be the turning point in the asbestos controversy in this country. In the case of CSR, Wittenoom presents a fear of the past returning to haunt the company for its brief flirtation with asbestos.

The questions which are at stake at Baryulgil and at Wittenoom ultimately concern the possibility of justice for a powerless and inarticulate group opposed by the intricate and powerful web of influences possessed by a major Australian company. In each instance, that company has enjoyed an intimate relationship with state and federal governments and has had ability to make appeal through publicity programmes designed to allay public fears. Both cases are a reminder of what can happen when the expectations of a workforce and its community are negligible. The absence of trade union influence, a critical public, and competent government authorities can so easily lead to a form of industrial anarchy.

There are numerous actors who have played a role in the asbestos drama, and in reviewing what happened at Wittenoom and Baryulgil many of the lesser players can easily be overlooked. The cast is headed by the asbestos workers and their families; then come the physicians who treated those men; then the companies and their executives who chose over a period of three decades to ignore the known hazards associated with asbestos fibre; there are lawyers who represent the victims and who also defend the corporations, in most cases, with marked success; there are the bureaucrats appointed to monitor and regulate the industry; there are the politicians at state and federal levels who are elected to identify and represent the public interest but who have, almost without exception, chosen to ignore the significance of the asbestos issue. Each of these actors occupies a specific place on an ascending ladder leading from the workplace to the state. And with each of the actors is found the same imperviousness to knowledge. The relationship of these multiple failures provides the intellectual history of asbestos. It is a story that has been played out at the quarry face, in the mill, the suburban factory, in the surgery, the boardroom, in courts, and in government offices. Many of the actors in this chain have sought, in various ways, to depoliticize the debate and to redraw the terms in which the debate is framed. Each of the actors also holds a particular model of what constitutes acceptable risk, and each possesses some notion of what is normal or acceptable for employees and consumers to expect of firms which produce hazardous products. The single most important group in this chain of participants are the physicians, for it is they who are the leading mediators of knowledge.

In the asbestos story, medical knowledge about disease has always been political knowledge. This is so despite the protestations of physicians, who claim that their work is immune from the influence of commerce or governments or interest groups. The way in which that knowledge is transposed into medical evidence against an industrial process, product or pollutant is vitally important in deciding the chances of justice faced by victims. With any hazardous substance there is a clear trade-off between work years and life expectancy on the one hand and the commercial and community benefits gained from the use of a particular product on the other. Lang Hancock's chilling comments on the benefits of asbestos to the Australian community contain an uncomfortable truth about the hidden assumptions which have operated in Australian industry for far too long.

Those assumptions are rarely, if ever, exposed to public gaze so that the community can decide if it wishes to endorse an excess number of deaths from lung cancer as the price to be paid for the benefits from a particular product or industrial process. In just the same way, environmental pollution is the cost the community pays for industry in terms of damaged landscape, stricken fauna and flora and unusable water. The two inventories are identical.

The assumptions about the nature of medical knowledge held by the industry and its critics and even by state authorities are much the same. Each assumes that knowledge grows in a linear fashion with past errors being relinquished, as new and better truths are established. That process is believed to be discreet and predictable. Where these three groups differ is in the selection of a particular aspect of the process of knowledge creation. In the Baryulgil controversy, James Hardie has chosen to emphasize the slowness of the emergence of proof linking asbestos with serious illness. The ALS, on the other hand, has sought to establish how early it was that knowledge emerged. Both sides base their case on the publication of specific research papers which are presumed to have established certain proof against asbestos. The ALS claims that the publication date, for instance, of the Merewether and Price report of 1930 should be the date from which the company's responsibility for asbestos exposure among its employees should be judged. In contrast, James Hardie has argued that the date of publication alone is no guide since, after that time, controversy about asbestos persisted for some years until the report was finally enthroned as a new ruling orthodoxy. And yet there is another date which both the industry and its critics ignore, and which has better claims for acceptance. That date begins with the existence of an oral tradition among physicians. Like Wagner's subsequent work on mesothelioma, the research by Merewether and Price was carried out over a period of years and, even when they first began work, there was already a suspicion among physicians treating asbestos workers that the fibre was hazardous and probably the cause of pneumoconiosis. In the case of Wagner's research on mesothelioma, Merewether's work on lung cancer had been published many years earlier, so there was good reason for Wagner's suspicions about the carcinogenic properties of the mineral.

If the asbestos industry is to be judged according to its professed concern for the health and welfare of employees and the users of its products, then the correct date for establishing responsibility must be set at the earliest days of oral tradition, and not from the date of seminal publications favoured by the industry's critics. The date of seminal publications is rarely the date at which a substance or a product was first suspected of being dangerous. In most cases the suspicion predated publication by some years and was the original reason for the research. It is on the basis of that original date that James Hardie, CSR, and the international conglomerates, Johns-Manville and Turner & Newall, should be called to account. Of course, if industry and its spokesmen had been less anxious to publicize the depth of their concern for the health of employees and their commitment to protect consumers it would not be fair to use the date of the oral tradition in assessing their responsibility. It is industry itself which has made such an approach necessary.

The dating of responsiblity is a vital question in the asbestos controversy, because of the proximity in time between medical knowledge about the dangers of the fibre and the massive expansion of the industry following World War II. If the date of knowledge about asbestos as a carcinogen is set before 1950 and not, as industry would have it, at some time after the publication of Selikoff's work in 1964, then many tens of thousands of lives are seen to have been wilfully thrown away. By employing this method, one can also judge dispassionately the notion of caution used by industry, as distinct from the hollow rhetoric which asbestos producers have paraded in their own defence. If by 1950 asbestos was known by physicians to be carcinogenic then the behaviour of companies which exposed workers to asbestos fibre is seen as contemptible. This method of dating also destroys the fallacy promoted to such effect by producers that industry has always been captive of imperfect medical knowledge, which trapped it in a web of ignorance and potential financial ruin, just as surely as it condemned generations of asbestos workers to a miserable death.

There is a second arena in which industry claims to be held captive by forces over which it has no control. Those forces are found in the technical process of production itself. Social philosophers, such as Langdon Winner and Ivan Illich,[2] suppose that high technology generates a momentum of its own which imposes a specific and narrow range of social and cultural choices upon its inventors and users. This idea has been taken up, albeit reflexively, by industry apologists, who claim that the technique of production enforces a kind of fate upon its users who, therefore, cannot be held responsible for the mistakes which the process precipitates. So it is with a sigh that spokesmen from James Hardie and Johns-Manville speak of the inadequacies of dust-monitoring equipment and ventilation systems, and the imperfections in plant and machinery which always generated more dust than owners wanted. This idea is implicit in much of the rhetoric used by corporations in their defence of industrial disaster. The idea of autonomous technology is never used systematically, and it appears as a kind of prevailing myth about the industrial process. It is a myth by which industry has sought to excuse itself of responsibility for the destruction of the health of employees and their families.

The notion of autonomous technology and the limits of medical knowledge each form part of the reason for industry's inertia in protecting workers from known hazards. Asbestos producers claim that they knew very little about the effects of fibre upon the human lung until the 1960s; that they did not have the means to monitor accurately the levels of fibre or dust; and that they did not possess the technology to eliminate dust from the workplace. Furthermore, if medical knowledge had been conclusive, so we are now told, producers would still have been forced to expose workers to fibre, or else Western economies would have had to give up using asbestos, which was such an invaluable and beneficial substance.

According to the industry's rhetoric, producers are as much the victims of asbestos as are the men and women whose health has been destroyed by pneumoconiosis or mesothelioma or lung cancer. It is somewhat bizarre that producers such as Johns-Manville now, more than anything, want public sympathy. Perhaps, even more than sympathy, producers wish to be excused from responsibility for the effects asbestos has had upon the health of individuals, families and entire communities.

The best means of defence against the kinds of events that occurred at Wittenoom and Baryulgil lies in a vigilant workforce and a sceptical public. As employees and as consumers, the public can control the behaviour and limit the excesses of producers by demanding effective regulation. There were ample legislative and regulatory mechanisms available at Wittenoom during the life of the Australian Blue Asbestos mine. If that legislation had been implemented few, if any, deaths would have occurred among ABA employees. The same is true of the Baryulgil mine, where the Department of Health and the Mines Inspectorate simply failed to carry out their responsibilities. In both instances there was adequate technology to reduce dust in the mines and mills; there were known means to protect workers in those areas where dust was inevitable; and there were long-established methods to judge the degree of risk. And yet the existence of these factors was insufficient to ensure that workers were protected. It is not sufficient to have regulations demanding certain kinds of behaviour from capital. It is necessary that governments can and do take action without being held captive by the very forces they are appointed to control. To regulate asbestos and other toxic substances requires greater supervision by the state as well as greater worker participation at the site of production. This, in turn, raises the problem about state intervention at a time when policies from both the right and the left in this country are demanding, in rhetoric at least, a diminution in the role of the state. A greater state presence is anathema to industry, except where it is confined to those specific actions which benefit industry directly. Wittenoom and Baryulgil document what the costs of following such a policy can be.

Wittenoom and Baryulgil are the first examples in Australia where the health of people other than employees has been affected by known industrial and environmental pollution. It is hoped that the suffering of the miners and their families will serve as sufficient warning, so that the experience will not be repeated. If not, that experience will not only occur again and again in factories and workshops but also among Australian consumers. Up until now, consumers in this country have been non-discriminating about the products they use and the guarantees of safety given so readily by producers. Perhaps the destruction of so many lives by asbestos fibre will, at last, force the community to realize that responsibility for the public interest must rest with the public itself and not with industry, no matter how respectable its voice may appear to be.

Notes to Chapters

Chapter 2

1. L. Noakes, *Asbestos* supplement, Department of Supply and Shipping, Mineral Resources of Australia, Summary Report No. 17, July 1945.
2. Ibid., p. 21, for full details of asbestos production in Australia in the period 1939–1944.
3. Ibid., p. 17.
4. Ibid., p. 7.
5. R.J. Hughes, "Asbestos in Australia — Its Occurrence and Resources", *Australian Mineral Industry Quarterly* 30 (1977), p. 122.
6. For the most comprehensive account of the history of the firm, see Geoffrey Hagan, "James Hardie Industries 1880–1980", B.A. (Hons.) thesis, Macquarie University, 1980.
7. Ibid., p. 32.
8. Tariff Board Report on Asbestos Fibre, 24 March 1955, Tariff Revision appendix B, p. 16.33.
9. Hagan, "James Hardie Industries 1880–1980", pp. 74-75.
10. Ibid., p. 75.
11. Ibid., p. 38.
12. Ibid., p. 36.
13. Ibid., p. 37.
14. John Reid, Chairman of James Hardie Asbestos Limited, in "Hardie Ferodo 1000: A James Hardie Group and Activity Report", 1978, p. 2.
15. James Hardie Asbestos Ltd, *Annual Report 1978,* p. 9.
16. Interhouse letter, James Hardie Head Office, Sydney, by Frank Page, March 1965.
17. James Hardie Asbestos Ltd, *Annual Report 1979,* p. 8.
18. James Kelso, James Hardie Service Industries, Evidence before the House of Representatives Standing Committee on Aboriginal Affairs Inquiry into *The Effects of Asbestos Mining on the Baryulgil Community,* transcript, December 1983, Sydney, p. 34.
19. Tariff Board Report on Asbestos Fibre, 24 March 1955, Tariff Revision appendix B, p. 16.
20. S.F. McCullagh, "The Biological Effects of Asbestos", *Medical Journal of Australia* 2 (1974), p. 48.
21. See "Prevention of Accidents: Safety Codes", in the *Report of the Working of the Factories and Shops Act 1912–46*, Sydney, 1949.

22. Cecil G. Roberts and Harry M. White, "A Survey of Dust Exposure and Lung Disease in the Asbestos Cement Industry in NSW", *Studies in Industrial Hygiene*, Division of Industrial Hygiene, Department of Health, NSW, n.d.
23. Ibid., p. 5.
24. Ibid., p. 9.
25. Ibid., p. 10.
26. See "Mining and Dust Trades", in Report of the Director General of Public Health, 1 January 1953 to 31 December 1957, Sydney, 1962, p. 93.
27. See NSW Joint Volumes of Papers, Legislative Council and Legislative Assembly, 1956–56, Vol. 3, 1957, p. 34.
28. See "Spraying of Asbestos", in Report of the Director General of Public Health for 1959, Sydney, 1961, p. 95.
29. *Economist*, 4 September 1982.
30. For an account of the internationalization of the asbestos industry, see Barry Castleman, "The Export of Hazardous Factories to Developing Nations", *International Journal of Health Services* 9, No. 4, 1979; and Castleman, "More on the International Asbestos Business", *International Journal of Health Services* 2, No. 2, 1981; also Kathleen Agena, "Hazards International: No Easy Solution is Possible", *The New York Times,* 27 April 1980.

Chapter 3

1. For an historical survey of pneumoconiosis see R. Sayers and A. Lanza, "A History of Silicosis and Asbestosis" in A.J. Lanza, *Silicosis and Asbestosis* (London: Oxford University Press, 1938), pp. 3-30.
2. See Annual Report of the Work of HM Women Inspectors of Factories, quoted in D. Lee and I. Selikoff, "Historical Background to the Asbestos Problem", *Environmental Research* 18 (1979), p. 300-314.
3. M. Anderson, "Historical Sketch of the Development of Legislation for Injurious and Dangerous Industries in England" in R. Oliver (ed.), *Dangerous Trades* (New York: Dutton, 1902).
4. W.E. Cooke, "Fibrosis of the Lungs due to the Inhalation of Asbestos Dust", *British Medical Journal* 2, 26 July 1924, p. 147.
5. These include T. Oliver, "Clinical Aspects of Pulmonary Asbestosis", *British Medical Journal* 3, 3 December 1927, pp. 1026-27; F.W. Simpson, "Pulmonary Asbestosis in South Africa", *British Medical Journal* 1, 26 May 1928, p. 885.
6. E.R.A. Merewether and C.W. Price, "Report on the Effects of Asbestos Dust on the Lungs and Dust Suppression in the Asbestos Industry: Part 1, Occurrence of Pulmonary Fibrosis and Other Pulmonary Afflictions in Asbestos Workers; Part 2, Processes Giving Rise to Dust and Methods for its Suppression" (London: HMSO, 1930).
7. Ibid., p. 7.
8. Ibid., p. 9.
9. Ibid., p. 10.
10. Ibid., p. 11.
11. Ibid., p. 15.
12. Ibid., p. 13.

13. Ibid., p. 28.
14. Ibid., p. 31.
15. "Memorandum on the Industrial Diseases of Silicosis and Asbestosis", Home Office (UK), July 1932.
16. A.J. Lanza, *Silicosis and Asbestosis* (London: Oxford University Press, 1938).
17. Ibid., p. 62.
18. Ibid., p. 245.
19. Ibid., p. 388.
20. Ibid., p. 4.
21. IARC Monographs on the Evaluation of the Carcinogenic Risk of Chemicals to Man, vol. 14 *Asbestos,* from IARC Working Group, Lyons, 14–17 December 1976, World Health Organization, p. 80.
22. Ibid., p. 81.
23. E.R.A. Merewether, *Annual Report of the Chief Inspector of Factories, 1947* (London: HMSO, 1949), pp. 68-81.
24. See Richard Doll, "Mortality from Lung Cancer in Asbestos Workers", *British Journal of Industrial Medicine* 12 (1955), pp. 81-86; S. Roodhouse Gloyne, "Pneumoconiosis: A Histological Survey of Necropsy Material in 1205 Cases", *Lancet* 1 (1951), pp. 810-14; L. Breslow et al., "Occupations and Cigarette Smoking as Factors in Lung Cancer", *American Journal of Public Health* 44 (1954), pp. 171-81.
25. See S. Shugar, "Effects of Asbestos in the Canadian Environment", National Research Council of Canada, Ottawa, NRCC/CNRC Publication, 1979; W.J. Nicholson et al., "Long Term Mortality Experience of Chrysotile Miners and Millers in Thetford Mines", *Annals of New York Academy of Science* 330 (1979), pp. 11-21.
26. Shugar, "Effects of Asbestos in the Canadian Environment", p. 124; see also M. Newhouse, "Epidemiology of Asbestos-Related Tumours", *Seminars in Oncology* 8, no. 3 (Sept. 1981).
27. Shugar, "Effects of Asbestos in the Canadian Environment", p. 22.
28. Ibid., p. 34.
29. IARC Monograph *Asbestos*, pp. 33-34.
30. Shugar, "Effects of Asbestos in the Canadian Environment", p. 52.
31. Ibid., p. 36.
32. Ibid., p. 42.
33. IARC Monograph *Asbestos,* p. 72.
34. J. Wagner, C.A. Sleggs and P. Marchand, "Diffuse Pleural Mesothelioma and Asbestos Exposure in the North-West Cape Province", *British Journal of Industrial Medicine* 17, no. 4 (1960), pp. 260-71.
35. Ibid., p. 262.
36. IARC Monograph *Asbestos*, p. 82.
37. Ibid., p. 84.
38. G. Ferguson and H. Watson, "Mesothelioma due to Domestic Exposure to Asbestos", *British Medical Journal* 288, 2 June 1984.
39. J.V. Sparks, "Pulmonary Asbestosis", *Medical Journal of Australia* 15 (March 1932), p. 388.
40. K.G. Outhred, "Pneumoconiosis on Western Australian Goldfields: An Outline of the Work of the Commonwealth Health Laboratory, Kalgoorlie", in *Collected Papers from the Third International Conference*

of Experts on Pneumoconiosis (Geneva: International Labour Office, 1950), pp. 184-204.

41. G. Smith and H.M. White, "An Investigation into the Incidence of Pneumoconiosis in NSW Coalminers and its Relation to Dust Exposure at Selected Colleries", ibid., pp. 205-62.
42. See discussion following Smith and White, ibid., pp. 231-32.
43. Dr Orenstein, ibid., p. 96.
44. Dr Vorwald, ibid., p. 91.
45. See Gordon M. Smith, "Occupational Factors in Pulmonary Dust Disease", *Medical Journal of Australia* 1, no. 22 (25 November 1950), pp. 777-82. Dr Smith also attended the conference in Sydney and took part in discussion on the subject of this article.
46. J.C. McNulty, "Malignant Pleural Mesothelioma in an Asbestos Worker", *Medical Journal of Australia* 2, 15 December 1962, pp. 953-54.
47. P.H. Kitto, "Dust and Dust Sampling in South Africa", *First Australian Pneumoconiosis Conference Proceedings* (University of Sydney, 12-14 February 1968), pp. 85-106.
48. Ibid., p. 93.
49. G. Major, "Dust Sampling: Instrumentation and Standards — A Review", ibid., pp. 7-44.
50. Ibid., p. 39.
51. M. Glick, "Pneumoconiosis in New South Wales Coal Miners", ibid., p. 177.
52. J. Cullen, "Pneumoconiosis at Broken Hill", ibid., pp. 237-77..
53. James Milne, "Fifteen Cases of Pleural Mesothelioma Associated with Occupational Exposure to Asbestos in Victoria", *Medical Journal of Australia* 2, no. 14 (1969), pp. 669-73.
54. I. Young, S. West, J. Jackson and P. Cantrells, "Prevalence of Asbestos Related Lung Disease Among Employees in Non-Asbestos Industries", *Medical Journal of Australia* no. 1 (1981), pp. 464-67.
55. Charles Mitchell, Brett Emmerson and Con Aroney, "Prevalence of Respiratory Morbidity in Brisbane Waterside Workers", *Medical Journal of Australia*, 8 August 1981, pp. 139-41.
56. Robert Barnes, "Compensatable Asbestos-Related Disease in New South Wales", *Medical Journal of Australia* 2, no. 5 (September 1984), pp. 221-24.
57. Ibid., p. 222.
58. William J. Nicholson, "Criteria Document for Swedish Occupational Standards: Asbestos and Inorganic Fibres", *Arbete Och. Halsa,* no. 17, 1981.
59. Shugar, "Effects of Asbestos in the Canadian Environment", p. 61.
60. See *The Royal College of Physicians* 16, no. 4 (October 1982), p. 213; for further discussion of the unreliability of death certification see N. Newhouse and J. Wagner, "Validation of Death Certificates in Asbestos Workers", *British Journal of Industrial Medicine,* no. 26, 1969, pp. 302-7; M. Rosenblatt, P. Teng and S. Kerpwe, "Diagnostic Accuracy in Cancer as Determined by Post-Mortem Examination", *Progress in Clinical Cancer,* no. 5, 1973, pp. 71-81. In the case of mesothelioma see M. Greenberg and R.A. Lloyd Davies, "Mesothelioma Register 1967-1968", *British Journal of Industrial Medicine,* no. 31, 1974, pp. 91-104, and M.L. Newhouse et al., "A Study of the Mortality of Female Asbestos Workers", *British Journal of Industrial Medicine,* no. 29, 1972, pp. 134-41.

61. J. McNulty, "Asbestos Mining, Wittenoom, Western Australia" in *First Australian Pneumoconiosis Conference Proceedings,* p. 451.
62. See Lanza, *Silicosis and Asbestosis* pp. 405-31.
63. Ibid., p. 410.
64. Ibid., p. 420.
65. For an account of the composition of the Committee and the process of the setting of the standard, see Richard Gillespie, "Risk Assessment in Occupational Health: The Asbestos Standard", Dept History and Philosophy of Science, University of Melbourne, 1978.
66. Ibid., p. 31.
67. *The Times* (London), 20 January 1975.
68. *Asbestos,* vol. 1, Final Report of the Advisory Committee, UK Health and Safety Commission, 1979.
69. Ibid., p. 80.
70. Ibid., appendix N, p. 88.
71. *Guardian* (London), 23 August 1983.

Chapter 4

1. Minutes and Votes and Proceedings of the Western Australian Parliament, 1st Session of 18th Parliament, 27 July–15 December 1944, vol. 2, p. 11.
2. Ibid., p. 30.
3. Ibid., p. 32.
4. Ibid., p. 42.
5. Minutes and Votes and Proceedings of the Western Australian Parliament, 2nd Session of 18th Parliament, 16 July–14 December 1945, vol. 2, p. 26.
6. Department of Mines file no. 789/45, sub-file no 1216/46, Subject Yampire Gorge–Wittenoom Gorge, 17/9/46, p. 115o.
7. Ibid., p. 115c.
8. Ibid., p. 115m.
9. Ibid.
10. Department of Mines file no. 321/43, letter from State Mining Engineer to Under Secretary of Mines, 7 March 1945, p. 143.
11. "Asbestos: World Production", *The West Australian,* 17 April 1945.
12. Department of Mines file no. 321/43, p. 145.
13. Keith Brown, evidence before Supreme Court of Western Australia no. 1052 of 1979 between Joan Mary Joosten, plaintiff, and Midalco Pty Ltd, defendant, transcripts, p. 374.
14. Department of Mines file no. 321/43, p. 135.
15. George King, evidence before Joosten trial, p. 322.
16. See Lenore Layman, "Occupational Health at Wittenoom, 1943–1966", paper presented to the ANZSERCH/APHA Conference, University of Adelaide, 20–24 May 1984.
17. Department of Mines file no. 789/45, letter from A.H. Telfer, Under Secretary of Mines, to the WA Manager CSR, 8 October 1946.
18. Department of Mines file no. 789/45, letter from Secretary of Department to Mr Powell, acting Director of ABA, 8 September 1947.
19. Cecil Broadhurst, evidence before Joosten trial, p. 430.

20. Ibid., p. 441.
21. Ibid., p. 444.
22. G. Major, "Asbestos Dust Exposure", in *The First Australian Pneumo-coniosis Conference Proceedings* (University of Sydney, 12–14 February 1968), p. 468.
23. Ibid., p. 470.
24. Ibid.
25. Ibid., p. 472.
26. J.C. McNulty, evidence before The House of Representatives Standing Committee on Aboriginal Affairs Inquiry into *The Effects of Asbestos Mining on the Baryulgil Community,* 21 February 1984, Perth, p. 1533.
27. Allan Osborne, evidence before Joosten trial, p. 250.
28. Ibid., p. 263.
29. Ibid., p. 268.
30. Ibid., p. 269.
31. George King, evidence before Joosten trial, p. 332.
32. Ibid., pp. 333-34.
33. Ibid., p. 365.
34. Ibid., p. 367.
35. Ibid., p. 371.
36. Ibid., p. 379.
37. "Basic Safety Rules", CSR pamphlet, n.d.
38. Question on notice from Mr Bryce to the minister for Health Western Australian Parliament, 24 March 1982, *Hansard*, p. 544.
39. Frederick Harmpell, interview, Perth, 3 October 1984.
40. Santo Janjetic, statement made at Perth, 10 July 1984.
41. Igino Casale, interview, Perth, 3 October 1984.
42. Zeff Penizza, interview, Perth, 3 October 1984.
43. Dennis Flowers and David Nobbs, interview, Perth, 5 October 1984.
44. Dennis Flowers, ibid.
45. David Nobbs, ibid.
46. Dennis Flowers, ibid.
47. J. Thomas interview, Perth, 4 October 1984.
48. J.C. McNulty, submission to the House of Representatives Standing Committee on Aboriginal Affairs Inquiry into *The Effects of Asbestos Mining on the Baryulgil Community* 21 February 1984, Perth, p. 1466.
49. Department of the North West file no. 35/59, p. 67.
50. Ibid., p. 63.
51. Department of the North West file no. 35/59, letter from ABA Manager O.A. Allan to the minister for Police, 30 September 1960.
52. Department of the North West file no. 35/59, letter from Arthur Griffith, minister for Housing, to the Manager, ABA, Wittenoom, p. 24.
53. Victor Kemp, interview, Perth, 6 October 1984.
54. Department of Public Health file no. 14/58, p. 13.
55. Ibid.
56. Department of Public Health, Inspector of Sanitation file no. 14/58, Tableland Shire Council Report.
57. Ibid., p. 123.
58. Ibid., p. 130.
59. J.C. McNulty, "Prevalence of Respiratory Symptoms in the Western Aust-

ralian Gold Miner compared with Coal Miners", *First Australian Pneumo-coniosis Conference Proceedings,* p. 466.

60. Lenore Layman, "Occupational Health at Wittenoom, 1943–1966", p. 4.
61. Question on notice, Mr Harman to the minister for Health, 9 November 1949, 2nd Session of 29th Parliament, *Hansard,* p. 4801.
62. Minute from Commission for Public Health, L. Henzell, to the minister for Public Health, 20 February 1950.
63. Letter from L. Henzell, Commissioner for Public Health, to Under Secretary of Mines, 26 October 1951.
64. Ibid.
65. The Western Australian Commission for Public Health *Annual Report 1959* p. 47.
66. Ibid., p. 48.
67. Ibid., p. 48.
68. Public Health file no. 729/52, letter from Deputy Commissioner for Public Health to Gordon Smith, Secretary, Committee on Industrial Hygiene, School of Public Health and Tropical Medicine, 11 September 1952.
69. Public Health file no. 473/64, 30 June 1953.
70. Ibid., p. 15.
71. Department of Mines *Annual Report 1945,* p. 18.
72. Department of Mines *Annual Report 1946,* p. 19.
73. Report from J.E. Lloyd, State Mining Engineer, Department of Mines, Kalgoorlie, 23 February 1948, "Australian Blue Asbestos — Wittenoom Gorge".
74. Ibid.
75. Report from J.E. Lloyd, State Mining Engineer, Department of Mines, Kalgoorlie, 26 November 1948, "Australian Blue Asbestos — Wittenoom Gorge".
76. Ibid., p. 2.
77. Ibid., p. 2.
78. Department of Mines *Annual Report 1948,* p. 24.
79. Department of Mines *Annual Report 1950,* p. 17.
80. Department of Mines file no. 31/52, "Wittenoom", 14 May 1952.
81. Department of Mines *Annual Report 1961,* p. 33.
82. John Faichney, evidence before the Joosten trial, p. 106.
83. Vincent Kemp and Dennis Flowers, interview, Perth, 10 October 1984.
84. Department of the North West file no. 5593/62, acc. no. 1540, letter from the minister for the North West to the Minister for Health, 19 March 1953.
85. Department of the North West file no. 5593/62, acc. no. 1540, "Wittenoom Gorge Hospital", letter from G. Oxer to the Commissioner for Public Health, 12 July 1963.
86. Department of Public Health file no. 5593/62, acc. no. 1540, memo Under Secretary of Public Health, 6 August 1963.
87. Lenore Layman, "Occupational Health at Wittenoom, 1943–1966", p. 14.
88. Vincent Kemp, interview, Perth, 7 October 1984.
89. *The Kalgoorlie Miner,* 13 May 1948.
90. Award: AWU (Blue Asbestos Works) No. 20/1945, *Western Australian Industrial Gazette,* 1950, vol. 27, pp. 47-54.
91. See Award Engineering (Blue Asbestos) No. 2 of 1953, Mining Blue Asbestos — AWU, *West Australian Industrial Gazette,* quarter ending 31 March 1957, vol. 37.

92. J.C. McNulty, evidence before the Baryulgil Inquiry 21 February, 1984, p. 1523.
93. Ibid., p. 1528.
94. Ibid., p. 1531.
95. *The Age*, 2 June 1979.

Chapter 5

1. Lenore Layman, "Occupational Health at Wittenoom, 1943–1966", paper presented to the ANZSERCH/APHA Conference, University of Adelaide, 20–24 May 1984, p. 14.
2. *The Age*, 15 June 1978.
3. S.F. McCullagh, "The Biological Effects of Asbestos", *Medical Journal of Australia* 2, 1974, pp. 45-49.
4. Nick D'Ascanio, Case 2: Asbestos Diseases Society submission to House of Representatives Standing Committee on Aboriginal Affairs Inquiry into *The Effects of Asbestos Mining on the Baryulgil Community*, 24 May 1984.
5. Ibid., p. 13.
6. Oscar Penetta, Case 3: Asbestos Diseases Society submission to The Baryulgil Inquiry, 24 May 1984.
7. Arthur Ballerum, Asbestos Diseases Society submission to The Baryulgil Inquiry, 24 May 1984.
8. Hubert Joosten, evidence before Supreme Court of Western Australia no. 1052 of 1979 between Joan Mary Joosten, plaintiff, and Midalco Pty Ltd, defendant, transcripts, p. 125.
9. Joan Joosten, evidence before Joosten trial, p. 42.
10. Hubert Joosten, evidence before Joosten trial, p. 130.
11. Patricia Lambert, Case 4: Asbestos Diseases Society submission to The Baryulgil Inquiry.
12. Julia Armstrong, Case 5: Asbestos Diseases Society submission to The Baryulgil Inquiry.
13. Janet Elder, "Asbestosis in Western Australia", *Medical Journal of Australia* 2, no. 13 (23 September 1967), pp. 579-83.
14. Ibid., p. 582.
15. Ibid., p. 582.
16. J.C. McNulty, "Asbestos Exposure in Australia", in *Proceedings of the International Pneumoconiosis Conference,* edited by H.A. Shapiro (London: Oxford University Press, 1980).
17. *The Health Hazard at Wittenoom,* Public Health Department, 8 December 1978.
18. Ibid., p. 1
19. Ibid., p. 4
20. Ibid., p. 4.
21. M.S.T. Hobbs, S.D. Woodward, B. Murphy, A.W. Musk, and J.E. Elder, "The Incidence of Pneumoconiosis, Mesothelioma and Other Respiratory Cancer in Men Engaged in Mining and Milling Crocidolite in Western Australia", in *Biological Effects of Mineral Fibres,* vol. 2, edited by J.C. Wagner (Lyons: IARC Scientific Publication no. 30, 1980), pp. 615-25.

22. Ibid., p. 624.
23. B. Armstrong, J. Hunt, C. Newall, and H. Henzell, Social and Preventative Medicine Project, "Epidemiology of Mesothelioma in Western Australia", reproduced in the transcripts of The Baryulgil Inquiry, 21 February 1984, Perth, pp. 1493-1521.
24. B. Armstrong et al., "Epidemiology of Malignant Mesothelioma in Western Australia", *Medical Journal of Australia* 141, no. 2 (21 July 1984), pp. 86-88.
25. Ibid., p. 88.
26. Question on notice from Hon. W. Olney to minister for Health, 30th Parliament of W.A., 1st Session, 1980.
27. *Report for the Year 1961*, WA Commission for Public Health, Perth, p. 44.
28. *Report for the Year 1962*, WA Commission for Public Health, Perth, p. 77.
29. *Report for the Year 1963*, WA Commission for Public Health, Perth, p. 66.
30. *Report for the Year 1966*, WA Commission for Public Health, Perth, p. 9.
31. Question on notice to minister for Health from Mr Hodge, WA 29th Parliament, 3rd Session, 21 November 1979, *Hansard*, pp. 5134-35. See also question on notice to mininster for Health, Mr Young, 29th Parliament, 2nd Session, 1978, *Hansard*, p. 3886.
32. *Annual Report 1982–83*, Workers Assistance Commission of Western Australia.
33. Brian Gandevia, "Comment: Mesothelioma", *Medical Journal of Australia* 141, no. 2 (21 July 1984), p. 78.
34. J.C. McNulty, "Asbestos Mining, Wittenoom Western Australia", 1st Australian Pneumoconiosis Conference, Sydney, February 1968, proceedings, p. 451.
35. See Question on notice from Hon. G. Berry to minister for Transport representing the Minister for Health, WA 29th Parliament, 2nd Session, 10 May 1978, *Hansard*, p. 1637.
36. *The Health Hazard at Wittenoom*.
37. *Ibid., p. 5.*
38. *The Age,* 5 December 1978.
39. *Exposure to Crocidolite at Wittenoom,* report prepared by Dr A. Cumpston, Public Health Department, Occupational Health Branch, 15 March 1979, reproduced in the transcripts of The Baryulgil Inquiry.
40. Ibid., p. 1450.
41. Ibid., p. 1454.
42. Joint news release from Government of WA and the Wittenoom Working Committee, 3 August 1979, p. 3.
43. Mr Davies, statement to the House, WA 29th Parliament, 3rd Session, 10 October 1979, *Hansard*, pp. 3479-83.
44. Mr Davies, debate on the future of Wittenoom, WA 29th Parliament, 3rd Session, 6 December 1979, *Hansard*, p. 6033-35.
45. *The Australian,* 3 October 1981.
46. *The West Australian,* 29 March 1982.
47. Question on notice Hon. Peter Dowling to the Chief Secretary, representing the minister for Health, 11 November 1982, WA 30th Parliament, 3rd Session, *Hansard*, p. 5123.

48. Question on notice Hon. Peter Dowling to the Chief Secretary representing the minister for Health, WA 16 November 1982, *Hansard*, pp. 5348-49.
49. *The Age*, 22 July 1983.
50. A.W. Musk, J.E. Baker, and D. Whitaker, "Sputum Asbestos Bodies and Radioligraphic Changes in Residents of Wittenoom, Western Australia", *Community Health Studies* 7, no. 1, 1983.
51. See submission by the Asbestos Disease Society to the House of Representatives Standing Committee on the Environment and Conservation, Inquiry into Hazardous Chemicals, 30 March 1981.
52. Ibid., *Hansard*, p. 897.

Chapter 6

1. Malcolm David Prentiss, "Aborigines and Europeans in the Northern Rivers Region of New South Wales, 1823–1881", M.A. thesis, Macquarie University, 1972.
2. George Farwell, *Squatters' Castle: The Story of a Pastoral Dynasty* (Melbourne: Lansdowne Press, 1973), p. 136; see also Malcolm Calley, "Bandjalang Social Organization", Ph.D. thesis, University of Sydney, for an account of the life of the people of the Baryulgil area.
3. Lucy Daley, interview, Baryulgil, 15 October 1984.
4. See the Annual Report of the Aborigines Welfare Board for the year ending 30 June 1944, Parliament of New South Wales, Appendix B, p. 19.
5. Summary of Report of the Department of Mines, 1954 (Second Session), Parliament of New South Wales (Sydney: Govt Printer, 1965), p. 9.
6. Ibid., p. 9.
7. Ken Gordon, interview, Baryulgil, 16 October 1984.
8. Submission by Hardie Trading (Services) Pty Ltd to the House of Representatives Inquiry into *The Effects of Asbestos Mining on the Baryulgil Community,* November 1983, p. 8.
9. "Mining Opportunity for North Coast: World's Largest Asbestos Field", *The Northern Star,* 22 June 1961.
10. Neil Walker, interview, Baryulgil, 22 October 1984.
11. Jerry Burke, evidence before the House of Representatives Standing Committee on Aboriginal Affairs Inquiry into *The Effects of Asbestos Mining on the Baryulgil Community,* Sydney, 7 February 1984, p. 123.
12. Submission by Jerry Burke to the Baryulgil Inquiry, 7 February 1983, Sydney, p. 2.
13. Jerry Burke, evidence before the Baryulgil Inquiry, Sydney, 7 February 1984, p. 122.
14. James Kelso, Hardie Trading (Services) Pty Ltd, evidence before the Baryulgil Inquiry, 2 December 1984, Sydney, p. 53.
15. Ibid., p. 57.
16. William Hindle, evidence before the Baryulgil Inquiry, 7 February 1984, Sydney, pp. 223-37.
17. Jerry Burke, evidence before the Baryulgil Inquiry, 7 February 1984, Sydney, p. 125.
18. Jerry Burke, ibid., p. 124.

19. James Kelso, evidence before the Baryulgil Inquiry, 2 May 1984, Canberra, p. 1627.
20. Hardie Papers Document 2: Letter Asbestos Mining Pty Ltd, 15 December 1966, untitled.
21. Jerry Burke, evidence before the Baryulgil Inquiry, 7 February 1984, Sydney, p. 126.
22. Jerry Burke, ibid., p. 134.
23. Jerry Burke, ibid., p. 134.
24. Rodney MacBeth, evidence before the Baryulgil Inquiry, 7 February 1984, Sydney, p. 215.
25. Greville Torrens, interview, Baryulgil, 20 October 1984.
26. Ken Gordon, interview, Baryulgil, 16 October 1984.
27. Ken Gordon, ibid.
28. Harrington, evidence before the Baryulgil Inquiry, 7 February 1984, Sydney, p. 109.
29. Charles Sheather, evidence before the Baryulgil Inquiry, 7 February 1984, Sydney, p. 247.
30. Neil Walker, interview, Baryulgil, 22 October 1984.
31. James Kelso, evidence before the Baryulgil Inquiry, 30 May 1984, Canberra, p. 1781.
32. "Asbestos and Aborigines", submission by Public Interest Advocacy Centre to the Baryulgil Inquiry, 9 February 1984, p. 623.
33. "Asbestos and Aborigines", p. 623.
34. "Safety in Industry", series of lectures delivered to inspectors of factories, Victoria, by K.A. Kinnish, Victorian Department of Labour, (Melbourne: Government Printer, 1954), p. 40.
35. Aboriginal Legal Service, submission part 3 to the Baryulgil Inquiry, 28 June 1984, Sydney, p. 2297.
36. "Asbestos and Aborigines", p. 679.
37. Ibid., p. 693.
38. Jerry Burke, evidence before the Baryulgil Inquiry, 7 February 1984, Sydney, p. 151.
39. "Asbestos and Aborigines", p. 693.
40. James Kelso, evidence before the Baryulgil Inquiry, 7 February 1984, Sydney, p. 333.
41. Ibid., p. 332.
42. Ibid., p. 351.
43. See "Membrane Filter Method for Estimating Airborne Asbestos Dust", NH & MRC, Australian Department of Health, October 1976, approved by the 82nd Session of Council.
44. "Exposure to Asbestos Dust in Crushing Plant and Quarry at Asbestos Mining Pty Ltd, Baryulgil", Scientific Division of Occupational Health, 24 May 1960, signed A.T. Jones, Scientific Officer, quoted in submission by New South Wales Department of Health, the Baryulgil Inquiry, 10 February 1984, Sydney.
45. Ibid., p. 877.
46. Ibid., p. 877.
47. Ibid., p. 875.
48. Document, Scientific Division of Occupational Health, 24 May 1960, signed A.T. Jones, Scientific Officer, quoted in submission by New South

Wales Department of Health, the Baryulgil Inquiry, 10 February 1984, Sydney.
49. For a summary of the official dust counts taken at the mine by the New South Wales Department of Health, see official *Hansard* report, the Baryulgil Inquiry, 10 February 1984, Sydney, p. 342.
50. "Reports of the Department of Mines, New South Wales, for the Year 1946" in Joint Volumes of Papers Presented to the Legislative Council and Legislative Assembly, vol. 3 (Sydney: Government Printer, 1950), p. 18.
51. Ibid., p. 27.
52. See "Report of the Working of the *Factories and Shops Act* 1912–1946", Department of Labour, Industry and Social Welfare, Parliament of New South Wales, Sydney, 1949.
53. Minute paper from Inspector of Mines titled "Inspection and Dust Sampling Asbestos Mining Pty Ltd", 1 February 1973, quoted in transcripts, the Baryulgil Inquiry, 10 February 1984, Sydney.
54. Minute paper Department of Mines, 19/12/73, subject Asbestos Mining Pty Ltd, Baryulgil inspection and dust sampling results, October 1973, signed Inspector of Mines (special duties).
55. Robert Marshall, evidence before the Baryulgil Inquiry, 10 February 1984, Sydney, p. 1068.
56. Evan Francis, evidence before the Baryulgil Inquiry, 10 February 1984, Sydney, p. 1071.
57. James Kelso, evidence before the Baryulgil Inquiry, 2 December 1983, Sydney, p. 24.
58. Robert Marshall, evidence before the Baryulgil Inquiry, 10 February 1984, Sydney, p. 1068.
59. James Kelso, evidence before the Baryulgil Inquiry, 10 February 1984, Sydney, *Hansard*, p. 42.1.
60. Hardie Papers, E.G. Reeve, Chief Draftsman, 26/2/66.
61. Ibid., p. 2.
62. Hardie Papers 89a, L.C. Denmead, Mine Manager's report for fortnight ending 1/4/69, dated 2/4/69.
63. Ibid.
64. Hardie Papers 5: "Report on the Industrial Hygiene Survey 14–17 September 1970", signed J. Winters, Industrial Hygiene Engineer, 16/10/70.
65. Hardie Papers 4: Interhouse letter to Manager, Baryulgil, from S.F. McCullagh, Medical Officer, subject Industrial Hygiene, Baryulgil, 6 November 1970.
66. Hardie Papers 20: Subject Department of Mines Inspection, 11 October 1973, to Head Office, from S.F. McCullagh, 21 February 1974.
67. Hardie Papers 13: Interhouse letter from S.F. McCullagh, subject Industrial Hygiene, Baryulgil, Medical Officer's Inspection (confidential), 7 February 1972.
68. Hardie Papers 47: Interhouse letter from S.F. McCullagh to Head Office, 29 February 1972, subject Dust Counts for February.
69. Hardie Papers 26: From J. Winters to Jerry Burke, subject Bi-Monthly Personal Dust Samples 21/9/76.
70. Jerry Burke, evidence before the Baryulgil Inquiry, 7 February 1984, Sydney, p. 152.
71. William Hindle, evidence before the Baryulgil Inquiry, 7 February 1984, Sydney, p. 224.

72. Ibid., p. 225.
73. Neil Walker, evidence before the Baryulgil Inquiry, 6 February 1984, Sydney, p. 96.
74. Neil Walker, interview, Baryulgil, 20 October 1984.
75. James Kelso, evidence before the Baryulgil Inquiry, 30 May 1984, Sydney, p. 1779.
76. Ibid., p. 1791.
77. James Kelso, evidence before the Baryulgil Inquiry, 2 December 1983,, Sydney, p. 66.
78. Hardie Papers 23: "Maintenance of Respirators", 23 February 1976.
79. Chris Lawrence on behalf of the Aboriginal Legal Service, evidence before the Baryulgil Inquiry, 13 December 1984, Sydney, pp. 49-51.
80. Quoted by James Kelso, evidence before the Baryulgil Inquiry, 2 December 1983, Sydney, p. 68.
81. Hardie Papers 46: Interhouse letter, March 1972.
82. William Hindle, evidence before the Baryulgil Inquiry, 7 February 1984, Sydney, p. 235.
83. Neil Walker, interview, Baryulgil, 20 October 1984.
84. James Kelso, evidence before the Baryulgil Inquiry, 10 February 1984, Sydney, pp. 461-71.
85. For a detailed account of the management of the mine during the Woodsreef period see David Barwick, Chairman and Managing Director of Woodsreef Mines Ltd, evidence and submission before the Baryulgil Inquiry, 28 June 1984, Sydney, pp. 2119-50.
86. Jerry Burke, evidence before the Baryulgil Inquiry, 7 February 1984, Sydney, p. 132.

Chapter 7

1. James Kelso, evidence before the House of Representatives Standing Committee on Aboriginal Affairs Inquiry into *The Effects of Asbestos Mining on the Baryulgil Community,* 2 December 1983, Sydney, p. 38.
2. James Kelso, evidence before the Baryulgil Inquiry, 10 February 1984,, Sydney, p. 36.2, official *Hansard* report.
3. James Kelso, evidence before the Baryulgil Inquiry, 2 December 1983, Sydney, p. 39.
4. Rodney MacBeth, evidence before the Baryulgil Inquiry, 6 February 1984, Sydney, p. 208.
5. Ibid., p. 210.
6. Donald Wilson, evidence before the Baryulgil Inquiry, 6 February 1984, Baryulgil, p. 117.
7. Ibid., pp. 117-18.
8. Neil Walker, interview, Baryulgil, 20 October 1984.
9. Ibid.
10. Letter from Warwick Sinclair to Cecil Patten, Aboriginal Legal Services, Redfern, 14 May 1983.
11. James Kelso, evidence before the Baryulgil Inquiry, 30 May 1984.
12. Ibid., p. 1752.

13. Pauline Gordon, interview, Baryulgil, 19 October 1984.
14. Ibid.
15. Chris Lawrence, evidence before the Baryulgil Inquiry, 2 December 1983, Sydney, p. 41.
16. Ibid., p. 41.
17. Ibid., p. 107.
18. Jerry Burke, evidence before the Baryulgil Inquiry, 7 February 1984, Sydney, p. 158.
19. Ibid., p. 159.
20. Charles Sheather, evidence before the Baryulgil Inquiry, 7 February 1984, Sydney, p. 247.
21. Pauline Gordon, evidence before the Baryulgil Inquiry, 6 February 1984, Sydney, p. 109.
22. James Kelso, evidence before the Baryulgil Inquiry, 2 December 1983, Sydney, p. 44.
23. James Kelso, statement before the Baryulgil Inquiry, ibid., p. 23.
24. Ibid., p. 24.
25. James Kelso, evidence before the Baryulgil Inquiry, 2 May 1984, Sydney, p. 1625.
26. See Submission by National Aboriginal Conference before the Baryulgil Inquiry, 7 February 1984, Sydney, esp. pp. 180-81.
27. K.C. Cross, "An Investigation of Degree of Asbestos Pollution in the Vicinity of Baryulgil and Bugilbar Gap, NSW", Document Prepared on Behalf of the Department of Aboriginal Affairs, 13 October 1980, p. 6.
28. Jerry Burke, evidence before the Baryulgil Inquiry, 7 February 1984, Sydney, p. 149.
29. Chris Lawrence, evidence before the Baryulgil Inquiry, 2 December 1984, Sydney, p. 102.
30. Cecil Patten, evidence before the Baryulgil Inquiry, 2 December 1984, Sydney, p. 102.
31. James Kelso, evidence before the Baryulgil Inquiry, 10 February 1984, Sydney, see *Hansard*, esp. pp. 36.3-37.2.
32. Death Certificate Cyril Mundine, cited in Baryulgil Inquiry, 10 February 1984, Sydney, *Hansard*, p. 37.1.
33. James Kelso, evidence before the Baryulgil Inquiry, 10 February 1984, Sydney, p. 37.4
34. Post-mortem Report Andrew Donnelly signed Dr K.D. Murray, 17/6/1977, Grafton Base Hospital, cited as Document Number 8a. Submission by Aboriginal Legal Service before the Baryulgil Inquiry (Yellow Books, Vol. 2).
35. Document, review of the Death Certificate of Andrew Donnelly, Dr R.J. Grobius, Grafton Base Hospital, 20 January 1984.
36. James Kelso, evidence before the Baryulgil Inquiry, 10 February 1984, Sydney, p. 39.2, official *Hansard* Report.
37. Radiology report of A. Preece by Dr A. Sharland, 15 January 1949, Grafton Base Hospital, cited in submission by the Aboriginal Legal Service before the Bryulgil Inquiry (Yellow Books, vol. 2, Document I).
38. Dr Pooks, radiology reports on A. Preece and H. Mundine, 2/4/52 (Yellow Books, Vol. 2, Document 2).
39. See submissions by the Aboriginal Legal Service before the Baryulgil In-

quiry, especially the Yellow Books, Vol. 1, Documents 11e and 11f. See also Submission by the Aboriginal Legal Service, 8 June 1984, "Medical Evidence of Morbidity from Asbestos-Related Diseases" before the Baryulgil Inquiry.

40. See Appendix A, ibid., p. 2188.
41. Ibid., pp. 2188-89.
42. Ibid., pp. 2195-2201.
43. Ibid., p. 2175.
44. "An Examination of the Aboriginal Miners of the Baryulgil Asbestos Mine", Respiratory Laboratory of the Division of Occupational Health, NSW, 1977.
45. "Investigation into the Health of the Baryulgil Asbestos Mine Workers", Division of Aboriginal Health, Division of Occupational Health Services Research, NSW, November 1979.
46. "A Re-Examination of the Health of the Miners and Ex-miners from the Baryulgil Asbestos Mine", Respiratory Laboratory of the Division of Occupational Health, NSW, 1981; "Further Examinations of the Baryulgil Asbestos Miners", Respiratory Laboratory of the Division of Occupational Health, NSW, 1982.
47. James Kelso, evidence before the Baryulgil Inquiry, 30 May 1984, Sydney, p. 1755.
48. G. Field, evidence before the Baryulgil Inquiry, 2 May 1984, Sydney, p. 1714.
49. Ibid., p. 1714.
50. Ibid., p. 1714.
51. "Aboriginal Health", Report to the minister for Health, L.J. Brereton, by the NSW Task Force for Aboriginal Health, September 1983.
52. A. Julienne, L. Smith, N. Thomson and A. Gray, "Summary of Aboriginal Mortality in New South Wales Country Regions, 1980–1981", NSW Department of Health, State Health Publication no. (IDS) 83-168, p. 9.
53. "Baryulgil", internal report from Division of Occupational Health, NSW Government, to the Commonwealth, 16 January 1981, reproduced in the Baryulgil Inquiry transcripts, 10 February 1984, pp. 884-85.
54. Correspondence between Chairman, Health Commission of NSW, to R. Walton, Aboriginal Health Branch Commonwealth Department of Health, 31 October 1979, reproduced in the Baryulgil Inquiry transcripts, 10 February 1984, p. 889.
55. Correspondence between Health Commission of NSW to R. Walton, Aboriginal Health Branch, Commonwealth Department of Health, 18 May 1979, reproduced in the Baryulgil Inquiry transcripts, 10 May 1984, p. 891.
56. *The Effects of Asbestos Mining on the Baryulgil Community,* final report of the House of Representatives Standing Committee on Aboriginal Affairs, October 1984, Canberra, p. 2.29.
57. Brownbill, Department of Aboriginal Affairs, evidence before the Baryulgil Inquiry, 30 May 1984, Sydney, p. 1824.
58. See transcripts the Baryulgil Inquiry, 30 May 1984, pp. 1820-40.
59. See Chapter 8, "Existing Legal Remedies" in *The Effects of Asbestos Mining on the Baryulgil Community.*
60. Chris Lawrence, Aboriginal Legal Service, evidence before the Baryulgil Inquiry, 14 December 1984, *Hansard,* p. 4.

61. Ibid., p. 23.
62. Jerry Burke, evidence before the Baryulgil Inquiry, 7 February 1984, Sydney, p. 151.
63. Ibid., p. 133.
64. Neil Walker, evidence before the Baryulgil Inquiry, 6 February 1984, Sydney, p. 100.
65. James Kelso, evidence before the Baryulgil Inquiry, 30 May 1984, Sydney, p. 1761.
66. Ibid., p. 1761.
67. Ibid., p. 1763.
68. James Kelso, evidence before the Baryulgil Inquiry, 2 December 1983, Sydney, p. 23.
69. James Kelso, evidence before the Baryulgil Inquiry, 30 May 1984, Sydney, p. 1612.

Chapter 8

1. James McNulty evidence before The House of Representatives Standing Committee on Aboriginal Affairs Inquiry into *The Effects of Asbestos Mining on the Baryulgil Community,* 21 February 1984, Perth, p. 1532.
2. Justice J. Wallace, Judgment Supreme Court of WA no. 1052 of 1979 between Joan Mary Joosten and Midalco Pty Ltd, 9 October 1979, p. 14.
3. *The West Australian,* 23 August 1978.
4. *The Age,* 2 June 1979.
5. Ibid.
6. *The West Australian,* 23 March 1983.
7. Rosemarie Vojakovic, interview, Perth, 21 October 1984.
8. M. Rees vs. Australian Blue Asbestos Pty Ltd, and B. O'Neill and W.A. Young, The Workers' Compensation Board of WA, Perth no. of matters 31/79.
9. Supreme Court of WA, 11-14 December 1981, Appeal No. 126 of 1981, Australian Blue Asbestos Pty Ltd, appellant, and Magnola Rees, Reasons for Judgment.
10. Laszlo Pataki, interview, Perth, 2 October 1984.
11. "A Submission Regarding Workers' Compensation in WA", prepared by the Asbestos Diseases Society of Australia Inc., October 1984.
12. Ibid., p. 2.
13. Asbestos Diseases Society, submission before the Baryulgil Inquiry, 24 May 1984.
14. *The Australian,* 1 December 1983.
15. *The Financial Review,* 14 August 1981.
16. *The Age,* 5 April 1984.
17. *Asbestos: Short Term Assistance,* Industries Assistance Commission Interim Report, 20 August 1978, No. 181.
18. Ibid., p. 18.
19. *Asbestos,* Industries Assistance Commission, 30 October 1979, No. 231, p. 4.

20. James Kelso, evidence before the Baryulgil Inquiry, 10 February 1984, Sydney, *Hansard*, p. 39.3.
21. S.F. McCullagh, "The Biological Effects of Asbestos", *Medical Journal of Australia*, July 1974.
22. N. Gilbert, discussion at the First Australian Pneumoconiosis Conference, Sydney, February 1968, collected papers, p. 505.
23. Ibid., p. 505.
24. Ibid., p. 505.
25. S.F. McCullagh, discussion at the First Australian Pneumoconiosis Conference, Sydney, 1968, p. 216.
26. Noel Arnold, *An Assessment of the Occupational Hygiene Programme at James Hardie and Company Ltd. Asbestos-cement Plants in Victoria, New South Wales and South Australia 1980-81,* Corporate Underwriting Group for The Federated Miscellaneous Workers Union of Australia, 1981.
27. Ibid., p. 46.
28. *The Australian,* 28 February 1979.
29. South Pacific Asbestos Association, submission to the Inquiry into Occupational Safety and Health, May 1980.
30. Ibid., p. 14.
31. Ibid., p. 1.
32. Max Austin, South Pacific Asbestos Association, evidence before the Inquiry into Occupational Safety and Health, 12 June 1980, p. 32.
33. Ibid., p. 36.
34. South Pacific Asbestos Association, 25 June 1981, presented before The House of Representatives Standing Committee on Environment and Conservation.
35. Ibid., p. 2.
36. Ibid., p. 13.
37. Ibid., pp. 13-14.
38. *Setting the Record Straight,* James Hardie pamphlet, n.d.
39. *Asbestos and Health: A Background Note,* James Hardie & Co Pty Ltd, February 1979.
40. W.A. Crawford, Director, Division of Occupational Health and Radiation Control, evidence before the Williams Inquiry, p. 28.
41. "James Hardie and Asbestos", *Business Review Weekly,* 12–15 November 1983, p. 12.
42. "The Purpose of Hardies Giant Rights Issue", *Rydges,* December 1983.
43. *The Age,* 20 November 1984.
44. *The Sydney Morning Herald,* 23 November 1984.
45. *The Effects of Asbestos Mining on the Baryulgil Community,* Report of the House of Representatives Standing Committee on Aboriginal Affairs, October 1984, Canberra, p. 1.24.
46. Ibid., p. 1.25.
47. Ibid.
48. James Kelso, evidence before the Baryulgil Inquiry, 2 December 1983, Sydney, p. 5.
49. Ibid., p. 7.
50. James Kelso, evidence before the Baryulgil Inquiry, 30 May 1984, Canberra, p. 1743.
51. Ibid., p. 1614.

52. *The Effects of Asbestos Mining on the Baryulgil Community,* p. 1.33.
53. Ian Cameron, the Baryulgil Inquiry, 14 December 1983, Sydney, p. 65.
54. Ibid., p. 67.
55. Ian Cameron, the House of Representatives Appropriation Bill, 13 October 1983, *Hansard,* p. 1766.
56. Ibid.
57. Ian Cameron, the House of Representatives, 9 October 1984, *Hansard,* p. 1293.
58. Jerry Hand, the House of Representatives, 4 October 1983, *Hansard,* p. 1277.
59. *The Effects of Asbestos Mining on the Baryulgil Community.*
60. Ibid., p. 6.28.
61. Ibid., p. 5.98.
62. Ibid., p. 1.31.
63. Ibid., p. 8.3.
64. Ibid., p. 7.22.
65. Ibid., p. 10.11.
66. Ibid., p. 10.26.
67. Ibid., p. 6.77; for a survey of the legislation in NSW see NSW Archives Authority, Guide to the State Archives: Record Group NSWCS Workers' Compensation (Silicosis) Committee: Preliminary inventory, Sydney 1965, NL 331 832 New, p. 7-12.
68. *The Effects of Asbestos Mining on the Baryulgil Community,* p. 9.25.
69. Dr Longley, evidence before the Baryulgil Inquiry, 10 February 1984, Sydney, p. 1158.
70. Ibid., p. 1167.
71. *Report of the Workers' Compensation (Dust Diseases) Board* for year ending 30 June 1981, NSW Parliament, p. 5.
72. *Report of the Workers' Compensation (Dust Diseases) Board* for year ending 30 June 1982, NSW Parliament, p. 5.
73. Report of Post-mortem, 5 September 1983, Karl Schultz, signed P. Bullpitt, the Prince of Wales Hospital, presented in the Yellow Books, vol. 2, ALS to the Baryulgil Inquiry.
74. Letter from Dr. I. Selikoff, the Mount Sinai Medical Centre, 28 March 1984, to Dr Lim, Trade Union Medical Centre, Yellow Books, vol. 2, ALS to the Baryulgil Inquiry.
75. G. Field, evidence before the Baryulgil Inquiry, 2 May 1984, Canberra, p. 1714.
76. Prof. Brian Gandevia, *Asbestos and Health in Perspective,* submission made on behalf of Woodsreef Mines Ltd, 20 July 1978, to the IAC Inquiry into the Barrabba Mines.
77. Ibid., pp. 380-81.
78. Ibid., p. 473.
79. Ibid., p. 382.
80. Prof. Brian Gandevia, evidence before the Baryulgil Inquiry, 28 June 1984, Sydney, p. 2239.
81. Ibid., p. 2278.

Chapter 9

1. A.L. Young, J.A. Caleagni, et al., "The Toxicology, Environmental Fate and Human Risk of Herbicide Orange and its Associated Dioxin", US Air Force (Occupational and Environmental Health Laboratory), Sir Brookes Force Base, October 978, p. vi-28.
2. See Richard Doll and Richard Peto, *The Causes of Cancer* (London: Oxford University Press, 1981).
3. See "Poison Cloud over Reagan", *The Sunday Times* (London), 6 March 1983.
4. T.G. Williams, *Report of the Commission of Inquiry into Occupational Health and Safety.*
5. Ibid., p. 33.
6. Appendix 7, "NSW Legislation Containing Provisions Relating to Occupational Safety and Health" in T.G. Williams, *Report,* p. 118.
7. Ibid., p. 45.
8. *The Washington Post,* 4 March 1983.
9. Brian Howe, House of Representatives, 9 December 1982, *Hansard,* p. 3302.
10. *Hazardous Chemicals,* House of Representatives Standing Committee on Environment and Conservation, Second Report No. 455/1982, December 1982.
11. *Hazardous Chemical Wastes: Storage, Transport and Disposal,* First Report of the Inquiry into Hazardous Chemicals, Report of the House of Representatives Standing Committee on Environment and Conservation, March 1982.
12. Ibid., p. 18.
13. *Hazardous Chemicals,* p. 76.
14. Ibid., p. 30.
15. Ibid., p. 113.
16. First Submission by Department of Science and the Environment, July 1980, before the House of Representatives Standing Committee on Environment and Conservation *Inquiry into the Management of Chemicals Hazardous to Health and the Environment.*
17. Ibid., p. 1.
18. Ibid., p. 2.
19. Ibid., p. 25.
20. Transcript, "Nationwide", ABC TV, 19 March 1979, p. 5.
21. Senator Don Grimes, minister representing the minister for Health, 17 November 1983, *Hansard,* p. 2813.
22. John Kerin, question on notice to the minister for the ACT, 20 February 1979.
23. John Kerin, question on notice to the minister for Housing and Construction, Mr Groom, 22 March 1979, *Hansard,* p. 1127.
24. Minister for Health, Mr Hunt, answer to question on notice No. 4065 from John Kerin, House of Representatives, *Hansard,* p. 2951.
25. Office of the Chairman, Capital Territory Health Commission, 17 November 1978, Public Statement No. 116.
26. ACT Health Commission, A/3641, from Dr Crowe, letter to Dr MacLeod, subject "Asbestos Insulation", 28/11/78.

27. Dr MacLeod's letter to the Chairman of the ACT Health Commission, 28 March 1979.
28. Document A: "Explanations for Estimates of Expenditure 1984/85", Department of Housing and Construction, circulated by the minister for Housing and Construction, the Hon. Chris Hyrford, August 1984.
29. Minister for Housing and Construction, Media Release, 21 August 1984.
30. *The Advertiser* (Adelaide), 6 November 1984.
31. H.G. Sforcina, Technical Manager, Architects Advisory Service, Royal Australian Institute of Architects, interview, 20 November 1984, Canberra.
32. Allan Vickers, Real Estate Institute of the ACT interview, 21 October 1984, Canberra.
33. McCann, Registrar, ACT Division of the Australian Institute of Valuers, interview, 22 October 1984, Canberra.
34. Submission by the Asbestos Diseases Society before the Inquiry into Hazardous Chemicals, 7 May 1980, prepared by Trevor Francis, House of Representatives Standing Committee on Environment and Conservation.
35. John Kerin, Grievance Debate, House of Representatives, 22 March 1979, *Hansard*, p. 1050.
36. *The Medical Aspects of the Effects of the Inhalation of Asbestos*, NH & MRC, adopted by 88th session of Council, October 1979.
37. *Report on the Health Hazards of Asbestos*, NH & MRC, prepared by the Asbestos Sub-Committee, and adopted by the 91st Session of the Council, June 1981, Commonwealth Department of Health.
38. Ibid., p. 4.
39. *Guideline on Asbestos*, ACTU/VHTC Occupational Health and Safety, October 1984.
40. *Asbestos Management*, Interim National Guide to The Protection of Workers from The Health Effect of Asbestos, National Consultative Committee on Occupational Health and Safety, June 1984, Department of Employment and Industrial Relations; *Controlling Asbestos Hazards*, Interim National Guide to Identification, Evaluation and Control of Asbestos Hazards, NCCOHS, June 1984.
41. Ibid., p. vii.

Chapter 10

1. *Congressional Quarterly: Almanac*, 97th Congress, 2nd session, 1982, vol. 38, p. 503.
2. See Landgon Winner, *Autonomous Technology* (Cambridge: M.I.T., 1977); and Ivan Illich, *Tools for Conviviality* (London: Calder & Boyars, 1973).

Selected Bibliography

Books, Theses, Monographs

Calley, Malcolm. "Bandjalang Social Organisation", Ph.D. thesis, University of Sydney, 1970.

Dalton, Alan J. *Asbestos Killer Dust: A Worker/Community Guide.* London: British Society for Social Responsibility in Science, 1979.

Doll, R., and R. Peto. *The Causes of Cancer.* London: Oxford University Press, 1981.

Doyal, Lesley (ed.). *Cancer in Britain: The Politics of Prevention.* London: Pluto Press, 1983.

Doyal, Lesley. *The Political Economy of Health.* London: Pluto Press, 1979.

Dubos, Rene. *Man, Medicine and Environment.* New York: Frederick A. Praegar, 1968.

Duffield, Robert. *Rogue Bull: The Story of Lang Hancock, King of Pilbara.* Sydney: Fontana/Collins, 1979.

Ehrenreich, John (ed.). *The Cultural Crisis of Modern Medicine.* New York: Monthly Review Press, 1978.

Farwell, George. *Squatters Castle: The Story of a Pastoral Dynasty.* Melbourne: Lansdowne Press, 1973.

Friedson, Elliot. *The Profession of Medicine.* New York: Dodd Mead, 1970.

Gillespie, Richard. "Risk Assessment in Occupational Health: The Asbestos Standard". B.A.(Hons.) thesis, University of Melbourne, 1978.

Gunningham, N. *Safeguarding the Worker: Job Hazards and the Role of Law.* Sydney: Law Book Co., 1984.

Hagan, Geoffrey. "James Hardie Industries 1880–1980". B.A. (Hons.) thesis, Macquarie University, 1980.

Hetzel, Basil. *Health and Australian Society.* Sydney: George Allen & Unwin, 1983.

Illich, Ivan. *Limits to Medicine.* London: Marion Boyars, 1976.

Krause, Elliot. *Power and Illness.* New York: Elsevier, 1977.

Lanza, A. *Silicosis and Asbestosis.* London: Oxford University Press, 1938.

McKeown, Thomas. *The Role of Medicine.* Princeton: Princeton University Press, 1979.

Moodie, P. *Aboriginal Health.* Canberra: ANU Press, 1973.

Navarro, Vincente. *Medicine under Capitalism.* London: Croom Helm, 1976.

Oliver, R. (ed.). *Dangerous Trades.* New York: Dutton, 1902.

Phillipson, Neill. *Man of Iron.* Melbourne: Wren Publishing, 1974.

Prentiss, Malcolm David. "Aborigines and Europeans in the Northern Rivers Region of New South Wales 1823–1881". M.A. thesis, Macquarie University, 1972.

Rowley, C.D. *Aboriginal Policy and Practice.* 3 vols. Canberra: ANU Press, 1970, 1971.

Selikoff, I.J., and D.H.K. Lee. *Asbestos and Disease.* New York: Academic Press, 1978.

Streeton, Hugh. *Capitalism, Socialism and the Environment.* London: Cambridge University Press, 1976.

World Health Organization. *Asbestos.* Vol. 14 of IARC Monographs on the Evaluation of the Carcinogenic Risk of Chemicals to Man. IARC Working Group, Lydon, 14–17 December 1974.

Articles

Armstrong, B., et al. "Epidemiology of Malignant Mesothelioma in Western Australia". *Medical Journal of Australia* 14 (2), 21 July 1984: 86-88.

"Baryulgil: The Story of a Public Health Disaster". *New Doctor* 35, March 1985: 20-23.

Cooke, W.E. "Fibrosis of the Lungs Due to the Inhalation of Asbestos Dust". *British Medical Journal* 2, 16 July 1924.

Doll, Richard. "Mortality from Lung Cancer in Asbestos Workers". *British Journal of Industrial Medicine* 12, 1955: 81-86.

Elder, J. "Asbestosis in Western Australia". *Medical Journal of Australia* 2 (13), 23 September 1969: 579-83.

Figlio, Karl. "Sinister Medicine? A Critique of Left Approaches to Medicine". *Radical Science Journal* 9, 1979: 14-68.

Hobbs, M.S.T., S.D. Woodward, B. Murphy, A.W. Musk, and J.E. Elder. "The Incidence of Pneumoconiosis, Mesothelioma and Other Respiratory Cancer in Men Engaged in Mining and Milling Crocidolite in Western Australia". In *Biological Effects of Mineral Fibres,* vol. 2, edited by J.C. Wagner. IARC Scientific Publication no. 30, 1980: 615-25.

Hughes, R.J. "Asbestos in Australia — Its Occurrence and Resources". *Australian Mineral Industry Quarterly* 30, 1977.

Layman, L. "Health at Wittenoom, 1943–1966". Paper presented to the ANZSERCH/APHA Conference, University of Adelaide, 20–24 May 1984.

Layman, L. "Work and Workers' Compensation Responses at Wittenoom, 1944–1966". *Community Health Studies* 3, 1983: 1-18.

McCullagh, S.F. "The Biological Effects of Asbestos". *Medical Journal of Australia* 2, 1974: 45-49.

McNulty, J.C. "Malignant Pleural Mesothelioma in an Asbestos Worker". *Medical Journal of Australia* 2, 15 December 1962: 953-54.

Milne, J. "Fifteen Cases of Pleural Mesothelioma Associated with Occupational Exposure to Asbestos in Victoria". *Medical Journal of Australia* 2 (14), 1969: 669-73.

Powles, John. "On the Limitations of Modern Medicine". *Science, Medicine and Man* 1, 1973: 1-30.

Stark, Evan. "What is Medicine:" *Radical Science Journal* 12, 1982: 48-89.

Selikoff, I. et al. "Asbestos Exposure and Neoplasia". *Journal of American Medical Association* 188, 1963: 142.

Vorwald, A.J. et al. "Asbestosis Studies Experimentally". Reprinted in *The Chemical Engineering and Mining Review,* 10 April 1951: 246.

Wagner, J., C.A. Sleggs, and P. Marchand. "Diffuse Pleural Mesothelioma and Asbestos Exposure in the North West Cape Province". *British Journal of Industrial Medicine* 17 (4), 1960: 260-71.

Reports, Submissions

Aboriginal Legal Service. Submissions before the House of Representatives Standing Committee on Aboriginal Affairs Inquiry into *The Effect of Asbestos Mining on the Baryulgil Community*, Parts 1, 2 and 3.

Asbestos Diseases Society of WA, Inc. "A Submission Calling for a Public Inquiry into the Needs of Asbestos Sufferers and Their Dependants". 2 July 1980.

––––––. "A Submission regarding Workers' Compensation in Western Australia". October 1984.

Australian Parliament. *The Effects of Asbestos Mining on the Baryulgil Community*. House of Representatives Standing Committee on Aboriginal Affairs transcripts, including submissions, proof and official Hansard reports.

––––––. *Hazardous Chemical Wastes: Storage, Transport and Disposal*. First report of the Inquiry into Hazardous Chemicals. Reports of the House of Representatives Standing Committee on Environment and Conservation, March 1982.

––––––. *Hazardous Chemicals*. Second report of the Inquiry into Hazardous Chemicals. Report of House of Representatives Standing Committee on Environment and Conservation, December 1982. Parliamentary Paper no. 455/1982.

Department of Mines. "Reports of the Department of Mines, New South Wales, for the years 1946–1949". In *Joint Volumes of Papers Presented to the Legislative Council and Legislative Assembly,* vol. 3. Sydney: Government Printer, 1950.

Dupre, J., J. Mustard, and R. Uffen. *Report of the Royal Commission on Matters of Health and Safety Arising from the Use of Asbestos in Ontario,* vol. 3. Toronto: Ontario Ministry of the Attorney General, 1984.

First Australian Pneumoconiosis Conference Proceedings. 2 vols. February 1968, Sydney.

Hardie Trading Services Pty Ltd. Submission to the House of Representatives Inquiry into the Effects of Asbestos Mining on the Baryulgil Community, November 1983.

Health and Safety Commission, UK. *Asbestos*, vol. 1. Final report of the Advisory Committee, 1979.

International labour Organization. *Collected Papers from the*

Third International Conference of Experts on Pneumoconiosis. Geneva, 1950.

Kinnish, K. A. "Safety in Industry". Series of lectures delivered to Inspectors of Factories, Victoria. Victorian Department of Labour, 1954.

Law Reform Commission of Western Australia. *Report on Limitation and Notice of Actions: Latent Disease and Injury.* Project no. 36, Part 1, October 1982.

Merewether, E.R.A. *Annual Report of the Chief Inspector of Factories, 1947.* London: HMSO, 1949, pp. 66-81.

Merewether, E.R.A., and C.W. Price. "Report on the Effects of Asbestos Dust on the Lungs and Dust Suppression in the Asbestos Industry: Part 1, Occurrence of Pulmonary Fibrosis and Other Pulmonary Affliction in Asbestos Workers; Part 2, Process Giving Rise to Dust and Method of Its Suppression". London: HMSO, 1930.

National Health and Medical Research Council. *Membrane Filter Method for Establishing Airborne Asbestos Dust.* Australian Department of Health, October 1976.

_____. *Report on the Health Hazards of Asbestos.* Canberra: AGPS, 1982.

National Occupational Health and Safety Commission. Interim Report, May 1984. Canberra: AGPS, 1984.

Osborne, P. *The Other Australia: The Crisis in Aboriginal Health.* Occasional Monograph 1, Department of Political Science, University of Tasmania, 1982.

Parliament of New South Wales. "Report of the Working Party of the Factories and Shops Act 1912–1946". Department of Labour Industry and Social Welfare, Sydney, 1949.

Public Health Department, Perth. *The Health Hazard at Wittenoom.* 8 December 1978.

Public Interest Advocacy Centre. "Asbestos and Aborigines". Submission before the House of Representatives Standing Committee on Aboriginal Affairs Inquiry into *The Effects of Asbestos Mining on the Baryulgil Community,* 9 February 1984.

Shugar, S. "Effects of Asbestos in the Canadian Environment". National Research Council of Canada, Ottawa. NRCC/CNRC Publication, 1979.

Smith, L.R. *Aboriginal Health Statistics in Australia: A Report and a Plan*. Health Research Project, Research Reports no. 2. Canberra: ANU Press, 1980.

United Kingdom. "Memorandum on the Industrial Diseases of Silicosis and Asbestosis". Home Office, July 1932.

Index

Aboriginal Affairs, Dept of, 179,
  180, 181
Aboriginal Legal Service, 142, 159,
  160, 165, 166, 168, 169, 171,
  184, 189, 216, 217, 218, 227, 261
Aboriginal Medical Service, 172
Adelaide, 238, 242
Allan, A.O., 115
Allan, Arthur, 156
Amalgamated Engineering Union,
  97
Amalgamated Metal Workers
  Union (AMWU), 20
Anthony, Doug, 204
Armstrong, Dr B., 116-17
Armstrong, Julia, 111, 112
Armstrong, Ronald, 111, 112
Arnold, Noel, 207
Asbestos
  Australian production of, 10, 11
  Qualities of, 9
  Uses of, 11, 12
  Varieties of exposure, 47
Asbestos disease
  Problems of diagnosis, 59, 60,
    114
  Problems in identifying
    causation, 60
  Litigation and, 109, 110, 111
  Quality of life and, 107-8, 112
Asbestos Diseases Society, 126-27,
  128, 129, 130, 193, 194, 197,
  198, 199, 201, 243
Asbestos Information
  Association of North America, 32
Asbestos International

Association, 32
Asbestosis
  Aetiology of disease, 42-43
  Cancer of the lung and, 43-44,
    45, 46
  Cigarette smoking and, 45, 49
  Growth of medical knowledge,
    45, 51, 60-62, 63
  History of disease, 36-42
  Other forms of cancer and, 46
Asbestos Mines Pty Ltd, 6, 10,
  149, 157, 158, 159, 164, 167,
  168, 170, 179, 183, 185, 187,
  220, 221 (*See also* James Hardie)
  Establishment of, 135
  Structure of company, 142
Asbestos Mining Company of
  Australia, 134
Austin, Max, 208-9, 210 (*See also*
  South Pacific Asbestos
  Association)
Australian Blue Asbestos, 5, 55, 71,
  72, 75, 76, 78, 79, 80, 81, 82, 85,
  86, 87, 88, 89, 93, 94, 95, 96, 97,
  99, 100, 101, 102, 103, 105-7,
  108, 109, 111, 112, 115, 117,
  119, 127, 190, 191, 192, 194,
  195, 200, 201, 202, 256, 264
Australian Capital Territory,
  243-50
Australian Council of Trade
  Unions, 252, 253
Australian Workers Union, 82, 97,
  98, 104, 140, 160, 161, 162, 192,
  196

Bagasse, 14
Bairnco Corporation (*See* Keene
    Corporation)
Baldwin-Ehret-Hill Incorporated,
    25, 26
Ballerum, Arthur, 104-5
Banjalang, 2, 131, 132, 133
Barnes, Robert, 58
Barraba, 158, 203, 208, 228
Baryulgil Mine
    Early operation of, 134-35
    Exposure of community, 164-65,
        167
    Community health surveys,
        173-75, 223
    Conditions in mill, 140,
        141
    Economic prospects of, 136
    Health survey at, 143, 156
    Living conditions at, 163-64
    Method of mining, 136-39
    Mines inspectorate role at, 144,
        147-49, 150, 154
    Wages paid at, 161
Baume, Sen. Peter, 180
Bayside Asbestos Removal, 4
Becklake, Dr M., 205
Better Brakes Pty Ltd, 17
Bhopal, 3, 4, 33, 186
Birmingham, 231
Bonegilla, 81
Brisbane, 57, 135, 238, 248
British Medical Journal, The, 38,
    50
British Occupational Hygiene
    Society, 66
Broadhurst, Cecil, 76
Broken Hill, 55, 56, 190
Brooklyn, 13
Burke, Jerry, 137, 139-40, 142,
    151, 153, 154, 157, 161, 165,
    168, 169, 171, 184, 186, 216

Camellia, 13, 15, 16, 17, 19, 143,
    147
Cameron, Ian, 217-18, 224
Canada, 9, 11, 22, 30, 47, 59,
    67, 74, 101, 135
Canberra, 7, 129, 181, 203, 243,
    244, 245, 247, 248, 250
Cape Asbestos Pty Ltd, 16, 67, 68

Caribbean, 33
Carter, Pres. J., 28, 32
Casale, Igino, 82
Casino, 136, 223
Cassair Asbestos Corporation, 16
Cellulose, 15
Charlemagne, 8
Chicago, 25, 27
China, Peoples Republic of, 31
Chrysolite Corporation of
    Australia, 203 (*See also*
    Woodsreef)
Clarence River, 131, 132, 134
Collum Collum Station, 179
Colonial Gorge, 76, 78
Colonial Sugar Refineries, 2, 4, 10,
    78
    History of, 14-15
    Involvement at Wittenoom, 70,
        73-74
COMECON, 31
Como, 107
Concord, 14
Cooke, W.E., 38
Copnanhurst Shire Council, 165
Court, C.W., 86
Crawford, Dr W., 212-13
Cross, K., 167
Cumpston, Dr, 121

Dales Gorge, 10
Daley, Mrs L., 132-33, 164
Darwin, 75, 81
D'Ascanio, Nick, 103
Davidson, W.S., 87
Denmark, 32
Dent, J.H., 67
Detroit, 231
Dieldrin, 186
Doctors Reform Society, 176
Doll, Sir Richard, 44
Donnelly, Andrew, 137, 169, 170,
    171, 173, 184, 226
Dow Chemical, 233
Dust Diseases Board (NSW), 19,
    58, 120, 146, 168, 169, 170, 175,
    176, 213, 225, 226, 227, 228,
    229, 252

Elder, Dr J., 112, 113
Electron Microscopy, 59, 60, 256

England, 13
Environment Protection Agency
  (USA), 230, 233, 241, 245
Europe, 9, 80
European Economic Community,
  32, 47, 67

Factory and Shops Act (NSW), 21
Faichney, John, 95
Farquhar Transport, 156
Federated Miscellaneous Workers
  Union, 19, 206
Fibrolite, 13
Field, Dr G., 175, 176, 227-28
Finland, 8
First Australian Pneumoconiosis
  Conference, 55
Florence, 132
Flowers, Dennis, 87
Flying Doctor Service, 96
Fortescue River, 125
France, 47, 60
Francis, Dr Eva, 150
Francis, Trevor, 126, 127, 129,
  243-44, 246, 248, 250, 251
Franklin, Benjamin, 9
Fremantle, 75, 81, 104

Gandevia, Prof. B., 120, 228, 229,
  235
Gilbert, N.E., 205-6
Gordon, Ken, 135, 140-41
Gordon, Pauline, 164, 166
Grafton, 131, 135, 136, 156, 163,
  164, 171, 180, 187
Grafton Base Hospital, 169
Greece, 36
Grimes, Sen. Don, 242
Grobius, Dr J., 169
Gundagai, 10
Gungahlin, 2

Hall, Timothy, 127
Hamburg, 132
Hamersley Ranges, 10
Hancock, Lang, 70, 71, 101-2, 122,
  124, 191, 260
Hand, Jerry, 216, 218-19
Hardie Ferodo Ltd, 17
Hardie Papers, 150-54, 157, 189,
  206, 216, 220

Hardiflex, 15
Harmpell, Frederick, 80, 81
Health Dept (WA), Public, 89-90
  Annual reports of, 91, 118, 125,
    118-19
  Studies of Wittenoom, 114-15,
    121-22
Hebden Bridge, 62
Henzell, Linley, 90, 92
Heroditus, 8
Hindle, William, 138-39, 154, 156
Holding, Clyde, 217
Holt, Sir Harold, 52
Housing and Public Works (NSW),
  Dept of, 14
Howe, Brian, 237
Hurford, Chris, 248

IARC, 43, 50, 229, 232, 238
Illich, Ivan, 263
ILO, 52, 67, 235
India, 3, 4, 31, 33
Indonesia, 17, 18, 35, 182, 211
Industries Assistance Commission,
  203-4, 228
Industrial Diseases Medical Board
  (*See* Pneumoconiosis Medical
  Board)
Isherwood, Joseph, 192
Italy, 3, 14, 31, 60, 80, 186, 255

Jakarta, 18, 34
James Hardie
  History of firm, 12-20
  Off-shore interests, 17, 18
  Processing of asbestos by, 15, 22
  Recruitment of employees, 13
  Retreat from asbestos products,
    34
  Technical innovations, 15, 16
Janjetic, Santo, 81
Japan, 3, 30, 31, 32, 46, 255
Joffre Creek, 125
Johns-Manville, 5, 9, 20, 24, 25,
  26, 27, 28, 29, 30, 62, 68, 69, 74,
  78, 129, 202, 207, 262, 263
Jones Creek, 10
Jones, R., 87
Joosten, Hubert, 106, 108
Joosten, Joan, 105-9, 250

Kalgoorlie, 52, 53, 90, 190, 196, 197
Keech Transport, 158
Keene Corporation, 25, 26, 27, 29
Kelso, James, 151, 155, 156, 167, 168, 170, 184, 185, 186, 205, 215, 216, 217
Kempsey, 136, 187
Kerin, John, 79, 80
Kitto, P.H., 55
Kiyembay, 31
Kuala Lumpur, 18
Kumbainggiri, 131

Labia, 9
Lambert, Harry, 110
Lambert, Patricia, 110, 111
Lanza, Dr J., 41, 42, 64, 65
Lawrence, Chris, 184, 218
"Liberty Ships", 28
Lidcombe Workers Health Centre, 246
Light Microscopy, 59
Little, W., 133
Lloyd, Charles, 141
Lloyd, J.E., 89, 93, 94
London, 9, 63
London Colosseum, 12
Longley, Dr E., 226
Love Canal, 186, 255
Lyon, 64

MacBeth, Rodney, 140, 160
McCullagh, Dr S.F., 21, 55, 102, 142, 153, 205, 207, 216
McKinney, John A., 25
McNulty, Dr J., 54, 55, 60, 85, 91, 99, 100, 112, 113, 114, 115, 118, 190
Malaysia, 17, 18, 34, 35, 182, 211
Manchester, 67
Manville Corporation (*See* Johns-Manville)
Maralinga, 186
Marra Mamba, 10
Marshall, Robert, 149
Medical Journal of Australia, The, 21, 52, 102, 205
Melbourne, 4, 13, 189, 231, 242, 252

Merewether, Dr, 23, 44, 54, 60, 62, 153, 205, 206, 207
Cancer of the lung and, 44-45
Merewether, Dr and Price, Dr, 38-39, 190, 220, 241
Recommendations made by, 39-40
Mesothelioma
Aeitology of disease, 48
Asbestos exposure and, 48-49, 50
Incidence in Australia, 54-55
Mexico, 33
Midalco (*See* Australian Blue Asbestos)
Middleton, Father H.S., 86
Midget Impinger, 59, 256
Minamata Bay, 3, 255
Milne, James, 57
Mines (WA), Dept of, 89-90, 92-93
Inspections of mine, 92-94, 95
Mines Ventilation Board, 98
Mitchell, Charles, 57
Monsanto, 233
Monte Bello Islands, 186
Montreal, 47
Mulabugilmah, 178, 179, 180, 181, 224
Muli Muli, 136, 223
Mundine, Cyril, 162, 168, 169, 171, 172, 225
Mundine, Harry, 171
Mundine, John, 161

National Aboriginal Conference, 167
National Health and Medical Research Council (NHMRC), 147, 150, 212, 238, 250, 251, 252
National Institute of Occupational Health and Safety, 65
National Library, 4, 7, 35, 247, 250
Netherlands, 60
New York, 20, 26, 37, 45, 64, 66, 191, 254
New York State, 3
New Zealand, 13, 17, 213
North America, 9
Northern Territory, 238
Norway, 8

Occupational Safety and Health
  Administration (OSHA), 65, 66
OEDC, 5, 31, 32, 233, 239
Ogilvie, Edward, 131, 132, 133
Oliver, Charlie, 98
Orenstein, Dr, 53
Osborne, Allan, 77, 78, 79, 94
Ottawa, 47
Outhred, K.C., 52
Oxer, Dr G., 96

Page, Sir Earle, 52
Page, Frank, 142, 155, 161
Panizza, Zeff, 82, 83
Paper making, 9
Papua New Guinea, 105
Parliament House, 2, 35, 245, 246
Parramatta, 13, 19, 242
Pataki, Laszlo, 196-97
Peacock, Mathew, 127, 170
Pearl Harbour, 75
Penetta, Oscar, 104, 105
Perth, 13, 15, 34, 70, 73, 74, 78,
  81, 82, 83, 84, 96, 98, 104, 105,
  111, 112, 129, 194, 197, 238
Phillipines, 33
Pilbara, 70, 71, 125
Pneumoconiosis Medical Board
  (WA), 111, 117, 119, 128, 195,
  196, 197, 198, 199
  Problems faced by, 199-200
Port Hedland, 96, 110
Port Sampson, 75, 104
Prentiss, M.D., 131
Public Health (NSW), Dept of, 22,
  23, 24
  Role at Baryulgil, 145-46
Public Interest Advocacy Centre,
  146, 159

QBE Insurance, 214-15
Quebec, 30
Queensland, 14

Reagan, Pres. R., 28, 32
Rees, Granville, 194, 195-96
Reeve, E.G., 152
Reid, Andrew, 12, 13
Reid, John, 19, 24
Rhodesia, 78
Riversdale, 13

Robertstown, 10
Rochdale, 67
Roeburne, 77, 95
Roma, 81
Royal Family, 1
Russia (*See* Soviet Union)

Saint, Dr Erich, 77
Sawer, Joseph, 99
Schultz, Karl, 227
Scotland, 9
Selikoff, Dr I., 26, 45, 60, 62, 210,
  227, 229
Seveso, 3, 186, 255
Sheather, Charles, 141, 166
Sinclair, Ian, 204
Sinclair, Warwick, 162, 163
Smith, Gordon, 53
Solfernio, 132
South Africa, 9, 10, 11, 22, 30, 31,
  47, 51, 54, 55, 60, 67, 74, 78,
  101, 134, 135
South Australia, 10
South Korea, 33
South Pacific Asbestos
  Association, 34, 208, 210
Soviet Union, 2, 10, 31
Spain, 31
Spark, J.V., 52
State Pollution Commission
  (NSW), 146
Supply and Shipping, Dept of, 11
Sunshine, 207, 241
Sweden, 32
Switzerland, 36
Sydney, 13, 14, 22, 55, 57, 58, 73,
  77, 104, 113, 142, 155, 161, 162,
  168, 182, 185, 189, 238
Sykes, D., 121
Szechuan province, 31

Tableland Shire Council, 88
Tabulum, 131, 136, 164
Taiwan, 33
Tariff Board, 14, 16, 17, 160
Tasmania, 10, 18, 70, 81, 206
Tasmanian Asbestos Pty Ltd, 10,
  14
Territories, Dept of, 2
Thailand, 31, 33
Thetford, 9, 62

Third International Conference on Experts on Pneumoconiosis, 52, 54
Times Beach, 236, 237
Tom Price, Mt, 110
Toronto, 47
Trade and Resources, Dept of, 18
Transvaal, 49
Turner & Newall (Turner Brothers), 9, 17, 18, 20, 29, 31, 62, 67, 207, 262

United Asbestos Cement Berhad of Malaysia, 18
United Kingdom, 9, 23, 24, 28, 39, 42, 62, 114
United States, 5, 21, 23, 24, 26, 27, 28, 29, 31, 32, 39, 44, 46, 47, 51, 54, 56, 59, 64, 65, 74, 78, 183, 204, 212, 214, 230, 233, 234, 236, 254, 255, 258

Vickers-Cockatoo Dockyards, 58
Vinyl chloride, 63, 232

Wagner, Dr J., 49, 50, 54, 55, 62, 115, 165, 261, 262
 Work on mesothelioma, 49-50, 89
Walker, Neil, 137, 141-42, 154, 155, 157, 161, 166, 184
Washington, 236
West Germany (FDR), 30, 31, 60, 233
Whim Creek, 75
Williams Inquiry, 126, 209, 211, 234-36
Wilson, Donald, 161
Winner, Langdon, 263
Winters, John, 55, 143, 152, 153, 206, 216

Wittenoom Gorge, 70, 71, 76, 90
Wittenoom Hotel, 71
Wittenoom mine
 Absence of physician at, 96
 Cost of living at, 83-84
 Dust allowance at, 97-99
 Housing shortage, 86-87
 Indebtedness of miners, 97
 Inspections of mine, 95
 Problems of dust control, 79, 103
 Problems of ore extraction, 77, 78
 Profitability of, 94-95, 101
 Processing of ore, 76
 Social structure of community, 84-85
 Town sanitation 87-88
 Water supply, 87
Wittenoom Trust, 191-92, 193, 194, 197
Wittenoom Working Committee, 123
Woodsreef Mines Pty Ltd, 135, 152, 153, 157, 158, 182, 203, 204, 224
Workers' Compensation and Assistance Act (WA), 197-98
Wunderlich, 14, 16, 18, 55, 134, 171, 191, 207, 211

Yamba, 136
Yampire Gorge, 10
Youammi, 75
Yugoslavia, 31, 81, 82
Yulgilbar, 131, 132, 133, 134, 135

Zeehan, 10, 14, 70, 73, 74